207 Current Topics in Microbiology and Immunology

Springer
Berlin
Heidelberg
New York
Barcelona
Budapest
Hong Kong
London
Milan
Paris
Santa Clara
Singapore
Tokyo

Prions Prions Prions

Edited by S.B. Prusiner

With 42 Figures

Springer

Stanley B. Prusiner, M.D.
Department of Neurology
HSE 781
University of California
San Francisco, CA 94143-0518
USA

Cover illustration: Molecular models of the cellular prion protein (left panel) and the scrapie protein (right panel). (Prepared by Zewei Huang and Fred Cohen.)

Cover design: Künkel+Lopka, Ilvesheim

ISSN 0070-217X

ISBN-13: 978-3-642-64640-9 e-ISBN-13: 978-3-642-60983-1
DOI: 10.1007/978-3-642-60983-1

Softcover reprint of the hardcover 1st edition 1996

Library of Congress Catalog Card Number 15-12910

Typesetting: Thomson Press (India) Ltd, Madras
SPIN: 10541812 27/3020/SPS – 5 4 3 2 1 – Printed on acid-free paper

List of Contents

List of Contributors

(Their addresses can be found at the beginning of their respective chapters.)

Human Prion Diseases and Neurodegeneration

S.B. PRUSINER

1 Introduction

Investigations of neurodegeneration in humans and animals caused by prions have produced many unexpected and striking findings. For many years three uncommon human diseases and several animal disorders were labeled transmissible encephalopathies, spongiform encephalopathies, or slow virus diseases (GAJDUSEK 1977, 1985; SIGURDSSON 1954). These illnesses are transmissible to experimental animals after a prolonged incubation period, and some features of the transmissible pathogen resemble those of viruses. However, early attempts to characterize the infectious pathogen causing scrapie of sheep and goats argue that these transmissible agents differ from both viruses and viroids (ALPER et al. 1967; HUNTER 1972).

A series of remarkable discoveries in the past three decades have led to the molecular and genetic characterization of the transmissible pathogen causing scrapie in animals and a quartet of human illnesses: kuru, Creutzfeldt-Jakob disease (CJD), Gerstmann-Sträussler-Scheinker disease (GSS), and fatal familial insomnia (FFI). To distinguish this pathogen from viruses and viroids, the term

Departments of Neurology and of Biochemistry and Biophysics, University of California, San Francisco, CA 94143, USA

"prion" was introduced to emphasize its proteinaceous and infectious nature (PRUSINER 1982).

An abnormal isoform of the prion protein, PrP^{Sc}, is the only known component of the prion (PRUSINER et al. 1981, 1984). PrP is encoded by a gene on the short arm of chromosome 20 in humans (SPARKES et al. 1986). PrP^{Sc} differs physically from the normal, cellular isoform PrP^{C} by its high β-sheet content, insolubility in detergents, propensity to aggregate, and relative resistance to proteolysis (MEYER et al. 1986; OESCH et al. 1985; PAN et al. 1993). While PrP^{Sc} is formed from PrP^{C} after the polypeptide chain is assembled (BORCHELT et al. 1990), attempts to identify a chemical modification that distinguishes the two PrP isoforms have been unsuccessful (STAHL et al. 1993). In contrast, biophysical studies have shown that PrP^{C} is rich in α-helices and is virtually devoid of β-sheet (PAN et al. 1993). The conversion of PrP^{C} into PrP^{Sc} appears to involve the unfolding of α-helices and their refolding into β-sheets. Since infectious prions are composed largely, if not entirely of PrP^{Sc}, this α-helix to β-sheet structural transition appears to be the fundamental event in the propagation of prions as well as the pathogenesis of neurodegeneration.

2 Clinical Manifestations of Prion Diseases

The human prion diseases are manifest as infectious, inherited, and sporadic disorders and are often referred to as kuru, CJD, GSS, and FFI depending upon the clinical and neuropathological findings (Table 1).

Infectious forms of prion diseases result from the horizontal transmission of the infectious prions, as occurs in iatrogenic CJD and kuru. Inherited forms, notably GSS, familial CJD, and FFI, comprise 10%–15% of all cases of prion disease. A mutation in the open reading frame (ORF) or protein coding region of the PrP gene has been found in all reported kindreds with inherited human prion disease. Sporadic forms of prion disease comprise most cases of CJD and possibly some cases of GSS (MASTERS et al. 1978). How prions arise in patients with sporadic forms is unknown but has been hypothesized to involve horizontal transmission, somatic mutation of the ORF of the PrP gene, and the spontaneous

Table 1. Human prion diseases

Disease	Etiology
Kuru	Infection
Creutzfeldt-Jakob disease	
Iatrogenic	Infection
Sporadic	Unknown
Familial	PrP mutation
Gerstmann-Sträussler-Scheinker disease	PrP mutation
Fatal familial insomnia	PrP mutation

conversion of PrP^{C} into PrP^{Sc} (GAJDUSEK 1977; PRUSINER 1989). Numerous attempts to establish an infectious link between sporadic CJD and a preexisting prion disease in animals or humans have been unrewarding (COUSENS et al. 1990; HARRIES-JONES et al. 1988; MALMGREN et al. 1979).

3 Diagnosis of Prion Diseases

A human prion disease should be considered in any patient who shows a progressive subacute or chronic decline in cognitive or motor function. Typically patients are adults between 40 and 70 years of age and often exhibit clinical features helpful in providing a premorbid diagnosis of prion disease, particularly sporadic CJD (BROWN et al. 1986). There is as yet no specific diagnostic test for prion disease in the cerebrospinal fluid. A definitive diagnosis of human prion disease, which is invariably fatal, can usually be made from the examination of brain tissue. Over the past 4 years knowledge of the molecular genetics of prion diseases has made it possible to diagnose inherited prion disease in living patients using DNA extracted from peripheral tissues.

The classic neuropathological features of human prion disease include spongiform degeneration, gliosis, and neuronal loss in the absence of an inflammatory reaction. When present, amyloid plaques which stain with α-PrP antibodies are diagnostic. The presence of protease-resistant PrP (PrP^{Sc} or PrP^{CJD}) in the infectious and sporadic forms and most of the inherited forms of these diseases implicates prions in their pathogenesis.

In some patients with inherited prion disease PrP^{Sc} is barely detectable or undetectable (BROWN et al. 1992; LITTLE et al. 1986; MANETTO et al. 1992; MEDORI et al. 1992), a situation mimicked in transgenic (Tg) mice which express a mutant PrP gene and spontaneously develop neurologic illness indistinguishable from experimental murine scrapie (HSIAO et al. 1990). Because molecular genetic analyses of PrP genes in patients with unusual dementing illnesses are readily performed, the diagnosis of inherited prion disease can often be established where there is either little or no neuropathology (COLLINGE et al. 1990), atypical neurodegenerative disease (MEDORI et al. 1992), or misdiagnosed neurodegenerative diseases (HESTON et al. 1966), including Alzheimer's disease. Although horizontal transmission of neurodegeneration to experimental hosts was for a time the "gold standard" of prion disease, it can no longer be used as such. Some investigators have reported that transmission of the inherited prion diseases from humans to experimental animals is frequently negative using rodents despite the presence of a pathogenic mutation in the PrP gene (TATEISHI et al. 1992) while others state that this is not the case with apes and monkeys as hosts (BROWN et al. 1993).

The hallmark common to all of the prion diseases whether sporadic, dominantly inherited, or acquired by infection is that they involve the aberrant metabolism of the prion protein (Prusiner 1991). Making a definitive diagnosis of human prion disease can be rapidly accomplished if PrP^{Sc} can be detected immunologically. In the familial forms of the prion diseases molecular genetic analyses of PrP can be diagnostic and performed on DNA extracted from blood leukocytes antemortem. Unfortunately, such testing is of little value in the diagnosis of the sporadic or infectious forms of prion disease.

In summary, the diagnosis of prion or prion protein disease may be made in patients on the basis of: (a) the presence of PrP^{Sc}, (b) mutant PrP genotype, or (c) appropriate immunohistology, and should not be excluded in patients with atypical neurodegenerative diseases until one or preferably two of these examinations have been performed (Collinge et al. 1990, 1992; Lantos et al. 1992).

4 Inherited Human Prion Diseases

Genetics were first thought to have a role in CJD with the recognition that approx. 10% of cases are familial (Davison and Rabiner 1940; Friede and DeJong 1964; Jacob et al. 1950; Kirschbaum 1924; Masters et al. 1979; Meggendorfer 1930; Rosenthal et al. 1976; Stender 1930). The discovery of the PrP gene and its linkage to scrapie incubation times in mice (Carlson et al. 1986) raised the possibility that mutation might feature in the hereditary human prion diseases. A proline (P) → leucine (L) mutation at codon 102 was shown to be linked genetically to development of GSS with a LOD score exceeding 3 (Fig. 1; Hsiao et al. 1989a). This mutation may be due to the deamination of a methylated deoxycytosine (C) coupled to deoxyguanosine (G) through a phosphodiester bond (CpG) in the germline DNA encoding PrP, which results in the substitution of deoxythymine (T) for deoxycytosine. The P102L mutation has been found in ten different families in nine different countries including the original GSS family (Doh-ura et al. 1989; Goldgaber et al. 1989; Goldhammer et al. 1993; Kretzschmar et al. 1991, 1992).

An insert of 144 bp at codon 53 containing six octarepeats was initially described in patients with CJD from four families all residing in southern England (Fig. 1; Collinge et al. 1992; Owen et al. 1989; Poulter et al. 1992). The mutation is thought to have arisen through a complex series of events since the human PrP gene contains only five octarepeats, indicating that a single recombination event could not have created the insert. Genealogic investigations have shown that all four families are related, arguing for a single founder born more than two centuries ago (Poulter et al. 1992). The LOD score for this extended pedigree exceeds 11. Studies from several laboratories have demonstrated that two, four, five, six, seven, eight, or nine octarepeats in addition to the normal five are found in individuals with inherited CJD (Brown et al. 1992; Goldfarb et al. 1991a; Owen et al. 1989, 1992) whereas deletion of one octarepeat has been identified without the neurologic disease (Laplanche et al. 1990; Palmer et al. 1993; Vnencak-Jones and Phillips 1992).

The unusually high incidence of CJD among Israeli Jews of Libyan origin was thought for many years to be due to the consumption of lightly cooked sheep brain or eyeballs (KAHANA et al. 1974, 1991). Recent studies have shown that some Libyan and Tunisian Jews in families with CJD have a PrP gene point mutation at codon 200 resulting in a glutamate (E) → lysine (K) substitution (GOLDFARB et al. 1990c; HSIAO et al. 1991a). One patient was homozygous for the E200K mutation, but her clinical presentation was similar to that of heterozygotes (HSIAO et al. 1991a), arguing that familial prion diseases are true autosomal dominant disorders. The E200K mutation has also been found in Slovaks originating from Orava in north-central Czechoslovakia (GOLDFARB et al. 1990c), in a cluster of familial cases in Chile (GOLDFARB et al. 1991b), in a large German family living in the United States (BERTONI et al. 1992) as well as in British (COLLINGE et al. 1993) and Japanese families (KITAMOTO 1993). Some investigators have argued that the E200K mutation originated in a Sephardic Jew whose descendants migrated from Spain and Portugal at the time of the inquisition (GOLDFARB et al. 1991b). It is more likely that the E200K mutation has arisen independently multiple times by the deamidation of a methylated CpG as described above the codon 102 mutation (HSIAO et al. 1989a, 1991a). In support of this hypothesis are historical records of Libyan and Tunisian Jews indicating that they are descended from Jews living on the island of Jerba where Jews first settled around 500 B.C. and not from Sephardim (UDOVITCH and VALENSI 1984).

Many families with CJD have been found to have a point mutation at codon 178 resulting in an aspartic acid (D) → asparagine (N) substitution (GOLDFARB et al. 1992a). In these patients as well as those with the E200K mutation PrP amyloid plaques are rare; the neuropathologic changes generally consist of widespread spongiform degeneration.

Insomnia has been described recently in multiple Italian families with the D178N mutation (MEDORI et al. 1992). The neuropathology in these patients with FFI is restricted to selected nuclei of the thalamus. It is unclear whether all patients with the D178N mutation or only a subset present with sleep disturbances; some of these patients have been labeled thalamic dementia (PETERSEN et al. 1992). DNA sequencing studies show that the allele with the D178N mutation encodes a methionine at position 129 in FFI while a valine is encoded at position 129 in familial CJD (GOLDFARB et al. 1992b). The discovery that FFI is an inherited prion disease clearly widens the clinical spectrum of these disorders and raises the possibility that many other degenerative diseases of unknown etiology may be caused by prions (MEDORI et al. 1992).

Like the E200K and D178N(V129) mutations, a valine (V) → isoleucine (I) mutation at PrP codon 210 produces CJD with classic symptoms and signs (POCCHIARI et al. 1993; RIPOLL et al. 1993). It appears that this V2101 mutation is also incompletely penetrant.

Other point mutations at codons 105, 117, 145, 198, 217, and possibly 232 also segregate with inherited prion diseases (DOH-URA et al. 1989; HSIAO et al. 1991b, 1992; KITAMOTO et al. 1993a,b). Patients with a dementing or telencephalic form of GSS have a mutation at codon 117. These patients as well as some in other families were once thought to have familial Alzheimer's disease but are now known to have prion diseases on the basis of PrP immunostaining of amyloid plaques and PrP gene mutations (GHETTI et al. 1989; GIACCONE et al. 1990; NOCHLIN et al. 1989). Patients with the codon 198 mutation have numerous neurofibrillary tangles that stain with antibodies to τ and have amyloid plaques (GHETTI et al. 1989; GIACCONE et al. 1990; NOCHLIN et al. 1989) that are composed largely of a PrP fragment extending from residue 58 to 150 (TAGLIAVINI et al. 1991). A genetic linkage study of this family produced a LOD score exceeding 6 (DLOUHY et al. 1992). The neuropathology of two patients of Swedish ancestry with the codon 217 mutation (IKEDA et al. 1991) was similar to that of patients with the codon 198 mutation.

Patients with GSS who have a substitution of leucine for proline at PrP codon 105 have been reported (KITAMOTO et al. 1993b). One patient with a prolonged neurologic illness spanning almost two decades with PrP amyloid plaques was found to have an amber mutation of the PrP gene resulting in a stop codon at residue 145 (KITAMOTO et al. 1993a). Staining of the plaques with α-PrP peptide antisera suggested that they might be composed exclusively of the truncated PrP molecules. That a PrP peptide ending at residue 145 polymerizes in amyloid filaments is to be expected since an earlier study noted above showed that the major PrP peptide in plaques from patients with the F198S mutation was an 11-kDa PrP peptide beginning at codon 58 and ending at about 150 (TAGLIAVINI et al. 1991). Furthermore, synthetic PrP peptides adjacent to and including residues 109–122 readily polymerize into rod-shaped structures with the tinctorial properties of amyloid (COME et al. 1993; FORLONI et al. 1993; GASSET et al. 1992; GOLDFARB et al. 1993).

5 Nomenclature for the Inherited Human Prion Diseases

Although each of the PrP mutations are associated with a typical clinical presentation as noted above, there are a sufficient number of exceptions that a particular mutation in a single pedigree can present with symptom complexes typical of CJD in some patients and GSS in others. Since we now know the molecular basis of the disorders, it seems preferable to name them according to the mutation and no longer refer to them as familial CJD, GSS, or FFI. Once the PrP gene mutation has been determined, we suggest that prion disease (P102L) be used in place of ataxic GSS such as that found in the original GSS family (GERSTMANN et al. 1936; KRETZSCHMAR et al. 1991), prion disease (E200K) instead of familial CJD in Libyan Jews, and prion disease (D178N, M129) instead of FFI (Table 2). These designations describe the precise etiologies of the disorders and remove any possible ambiguities.

The need to designate the inherited prion diseases by their mutations (molecular lesions) is emphasized by the vastly different clinical presentations and postmortem neuropathologies observed in four afflicted members of a family with

Table 2. Inherited prion diseases of humans

Proposed designation	Alternative name
Inherited prion disease (P102L)	GSS
Inherited prion disease (P105L)	GSS
Inherited prion disease (A117V)	GSS
Inherited prion disease (Y145Stop)	GSS
Inherited prion disease (D178N)	Familial CJD, FFI
Inherited prion disease (V1801)	GSS
Inherited prion disease (F198S)	GSS
Inherited prion disease (E200K)	Familial CJD
Inherited prion disease (V210I)	Familial CJD
Inherited prion disease (Q217R)	GSS
Inherited prion disease (octarepeat insert)	Familial CJD

prion disease (6 octarepeat insert) (COLLINGE et al. 1992). One of the four family members with the insert presented a classical case of CJD and had pronounced spongiform change in the cerebral cortex while another presented with ataxia and had numerous PrP amyloid plaques at autopsy. The second case might have been called GSS with hesitation. The third and fourth members of the family died in hospitals with the diagnosis of dementia but had no spongiform change at autopsy and were not given the diagnosis of CJD.

6 Human PrP Gene Polymorphisms

At PrP codon 129, a polymorphism encodes either methionine (M) or valine (V) (Fig. 1, OWNE et al. 1990a). In whites this polymorphism appears able to influence prion disease expression not only in inherited forms but also in iatrogenic and sporadic forms of prion disease. A second polymorphism resulting in an amino acid substitution at codon 219 (E → K) has been reported in the Japanese population, in which the K allele occurs with a frequency of 6% (KITAMOTO and TATEISHI 1994).

Susceptibility to infection may be partially determined by the PrP codon 129 genotype (COLLINGE and PALMER 1991), analogous in principle to the incubation-time alleles in mice (CARLSON et al. 1986; COLLINGE and PALMER 1991). Population frequencies for the codon 129 polymorphism in whites are 12% V/V, 37% M/M, and 51% M/V (PALMER et al. 1991). In 16 patients (15 white, 1 black) from the United Kingdom, United States and France with iatrogenic CJD from contaminated growth hormone extacts, eight (50%) were V/V, five (31%) were M/M, and three (19%) were M/V (COLLINGE and PALMER 1991). Thus, a disproportionate number of patients with iatrogenic CJD were homozygous for valine at PrP codon 129. In contrast, the frequency of the V allele in the Japanese population is much lower (DOH-URA et al. 1991; MIYAZONO et al. 1992), and heterozygosity at codon 129 (M/V) is more frequent (18%) in CJD patients than the general population where

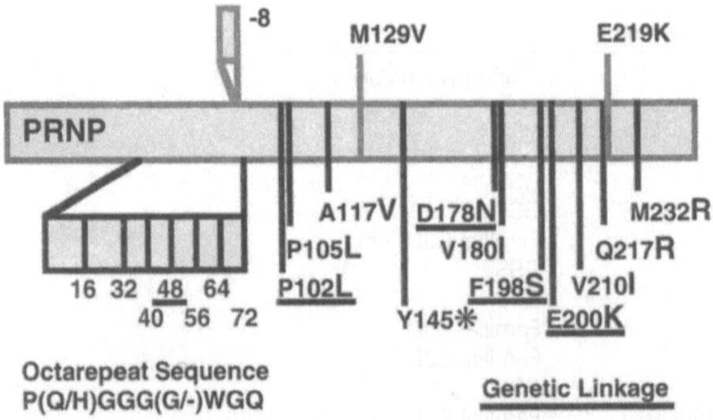

Fig. 1. Human prion protein gene (*PRNP*). *Large gray rectangle*, ORF; *above*, human PRNP wild-type polymorphisms; *below*, mutations that segregate with the inherited prion diseases. The wild-type human PrP gene contains five octarepeats [P(Q/H)GGG(G/-)WGQ] from codon 51 to 91 (KRETZSCHMAR et al. 1986). Deletion of a single octarepeat at codon 81 or 82 is not associated with prion disease (LAPLANCHE et al. 1990; PUCKETT et al. 1991; VNENCAK-JONES and PHILLIPS 1992); whether this deletion alters the phenotypic characteristics of a prion disease is unknown. There are common polymorphisms at codons 117 (Ala → Ala) and 129 (Met → Val); homozygosity for Met or Val at codon 129 appears to increase susceptibility to sporadic CJD (PALMER et al. 1991). *Small rectangle below ORF*, octarepeat inserts of 16, 32, 40, 48, 56, 64, and 72 amino acids at codons 67, 75, or 83. These inserts segregate with familial CJD and significant genetic linkage has been demonstrated where sufficient specimens from family members are available (COLLINGE et al. 1989, 1990; CROW et al. 1990; GOLDFARB et al. 1990b, 1991a; OWEN et al. 1989, 1990b; PALMER et al. 1993). Point mutations are designated by the wild-type amino acid preceding the codon number and the mutant residue follows, i.e., P102L. These point mutations segregate with the inherited prion diseases, and significant genetic linkage (*underlined mutations*) has been demonstrated where sufficient specimens from family members are available. Mutations at codons 102 (Pro → Leu), 117 (Ala → Val), 198 (Phe → Ser), and 217 (Gln → Arg) are found in patients with GSS (DOH-URA et al. 1989; GOLDFARB et al. 1990a–c; GOLDGABER et al. 1989; HSIAO et al. 1989a,b, 1991b; HSIAO and PRUSINER 1990; TATEISHI et al. 1990). Point mutations at codons 178 (Asp → Asn), 200 (Glu → Lys), and 210 (Val → Iso) are found in patients with familial CJD (GABIZON et al. 1991; GOLDFARB et al. 1990c, 1991c; HSIAO et al. 1991a; RIPOLL et al. 1993). Point mutations at codons 198 (Phe → Ser) and 217 (Gin → Arg) are found in patients with GSS who have PrP amyloid plaques and neurofibrillary tangles (DLOUHY et al. 1992; HSIAO et al. 1992). Additional point mutations at codons 145 (Tyr → Stop), 105 (Pro → Leu), 180 (Val → Iso), and 232 (Met → Arg) have been recently reported (KITAMOTO et al. 1993a,b). Single letter code for amino acids is as follows: *A*, Ala; *D*, Asp; *E*, Glu; *F*, Phe; *I*, Iso; *K*, Lys; *L*, Leu; *M*, Met; *N*, Asn; *P*, Pro; *Q*, Gln; *R*, Arg; *S*, Ser; *T*, Thr; *V*, Val; and *Y*, Tyr. (Reproduced with permission from PRUSINER 1993)

the polymorphism frequencies are 0% V/V, 92% M/M, and 8% M/V (TATEISHI and KITAMOTO 1993). Whether these associations are significant awaits statistical analysis of larger samples.

No specific mutations have been identified in the PrP gene of patients with sporadic CJD. However, patients with sporadic CJD are largely homozygous at codon 129 (PALMER et al. 1991). This finding supports a model of prion production which favors PrP interactions between homologous proteins, as appears to occur in Tg mice expressing Syrian hamster (SHa) PrP inoculated with either hamster prions or mouse prions (PRUSINER 1991; PRUSINER et al. 1990; SCOTT et al. 1989) as well as Tg mice expressing a chimeric mouse/hamster PrP transgene inoculated with "artificial" prions (SCOTT et al. 1993).

Approximately 15% of patients with sporadic CJD develop ataxia as an early sign, accompanied by dementia (BROWN et al. 1984). Most but not all patients with ataxia have compact (kuru) plaques in the cerebellum (PEARLMAN et al. 1988). Patients with ataxia and compact plaques exhibit a protracted clinical course which may last up to 3 years. The molecular basis for the differences between CJD of shorter and longer duration have not yet been fully elucidated. However, some preliminary analyses have suggested that patients with protracted, atypical clinical courses are more likely to be heterozygous at codon 129 (COLLINGE and PALMER 1991; DOH-URA et al. 1991).

Homozygosity at codon 129 has been reported to be associated with an earlier age of onset in the inherited prion disease caused by the 6 octarepeat insert but not by the E200K mutation in Libyan Jews (BAKER et al. 1991; GABIZON et al. 1993). As noted above, the FFI phenotype is found in patients with the D178N mutation who encode a methionine at codon 129 on the mutant allele while those with dementing illness (familial CJD) encode a valine at 129 (GOLDFARB et al. 1992b). Homozygosity for either M or V at codon 129 is thought to be associated with an earlier age of onset for the D178N mutation.

7 Barriers for the Transmission of Prion Diseases

Passage of prions between species is a stochastic process characterized by prolonged incubation times (PATTISON 1966). Prions synthesized de novo reflect the sequence of the host PrP gene and not that of the PrP^{Sc} molecules in the inoculum (BOCKMAN et al. 1987). On subsequent passage in a homologous host the incubation time shortens to that recorded for all subsequent passages, and it becomes a nonstochastic process. The species barrier concept is of practical importance in assessing the risk for humans of developing CJD after consumption of scrapie-infected lamb or from cattle with bovine spongiform encephalopathy (PRUSINER et al. 1993a; WILESMITH et al. 1992).

To test the hypothesis that differences in PrP gene sequences are responsible for the species barrier, Tg mice expressing SHaPrP were constructed (PRUSINER et al. 1990; SCOTT et al. 1989). The PrP genes of Syrian hamsters and mice encode proteins differing at 16 positions. Incubation times in four lines of Tg(SHaPrP) mice inoculated with Mo prions were prolonged compared to those observed for non-Tg, control mice. Inoculation of Tg(SHaPrP) mice with SHa prions demonstrated abrogation of the species barrier resulting in abbreviated incubation times due to a nonstochastic process (PRUSINER et al. 1990; SCOTT et al. 1989). The length of the incubation time after inoculation with SHa prions was inversely proportional to the level of $SHaPrP^{C}$ in the brains of Tg(SHaPrP) mice (PRUSINER et al. 1990). $SHaPrP^{Sc}$ levels in the brains of clinically ill mice were similar in all four Tg(SHaPrP) lines inoculated with SHa prions. Bioassays of brain extracts from clinically ill Tg(SHaPrP) mice inoculated with Mo prions revealed that only Mo prions but no SHa prions were produced. Conversely, inoculation of Tg(SHaPrP) mice with SHa prions led to only the synthesis of SHa prions. These findings contend that the amino acid sequence of PrP^{Sc} in the inoculum specifies whether Mo or $SHaPrP^{C}$ is converted into PrP^{Sc} in recipient Tg(SHaPrP) mice. Furthermore, the production of PrP^{Sc} appears to involve the formation of an intermediate complex between homotypic PrP^{C} and PrP^{Sc}.

The neuropathology of Tg(SHaPrP) mice was also found to be determined by the genetic origin of prion inoculum. Mo prions injected into Tg(SHaPrP) mice produced a neuropathology characteristic of mice with scrapie. A moderate degree of vacuolation in both the gray and white matter was found while amyloid plaques were rarely detected. Inoculation of Tg(SHaPrP) mice with SHa prions produced intense vacuolation of the gray matter, sparing of the white matter and numerous SHaPrP amyloid plaques characteristic of Syrian hamsters with scrapie.

8 Transgenic Mice Overexpressing Wild-Type PrP Genes

During transgenetic studies we discovered that uninoculated older mice harboring high copy numbers of wild-type (wt) PrP transgenes derived from Syrian hamsters, sheep, and PrP-B mice spontaneously developed truncal ataxia, hind-limb paralysis, and tremors (Westaway et al. 1994). These Tg mice exhibited a profound necrotizing myopathy involving skeletal muscle, a demyelinating polyneuropathy, and focal vacuolation of the CNS. Development of disease was dependent on transgene dosage. For example, Tg($SHaPrP^{+/+}$)7 mice homozygous for the SHaPrP transgene array regularly developed disease between 400 and 600 days of age while hemizygous Tg($SHaPrP^{+/0}$)7 mice also developed disease but only after 650 days.

Attempts to demonstrate PrP^{Sc} in either muscle or brain were unsuccessful, but transmission of disease with brain extracts from Tg($SHaPrP^{+/+}$)7 mice inoculated into Syrian hamsters did occur. These Syrian hamsters had PrP^{Sc} as detected by immunoblotting and spongiform degeneration (D. Groth and S.B. Prusiner, unpublished data). Serial passage with brain extracts from these animals to recipients was observed. De novo synthesis of prions in Tg($SHaPrP^{+/+}$)7 mice overexpressing $wtSHaPrP^{C}$ provide supports for the hypothesis that sporadic CJD does not result from infection but rather is a consequence of the spontaneous, although rare, conversion of PrP^{C} into PrP^{Sc}. Alternatively, a somatic mutation in which mutant $SHaPrP^{C}$ is spontaneously converted into PrP^{Sc} as in the inherited prion diseases could also explain sporadic CJD. These findings as well as those described above for Tg(MoPrP-P101L) mice argue that prions are devoid of foreign nucleic acid, in accord with many earlier studies that use other experimental approaches (Prusiner 1991).

9 Conclusions and Prospective

Knowledge accrued from the study of prion diseases may provide an effective strategy for defining the etiologies and dissecting the molecular pathogenesis of the more common neurodegenerative disorders such as Alzheimer's disease, Parkinson's disease, and amyotrophic lateral sclerosis (ALS). Advances in the molecular genetics of Alzheimer's disease and ALS suggest that, as in the prion diseases, an important subset are caused by mutations that result in nonconservative amino acid substitutions in proteins expressed in the CNS.

At present there are no effective therapies for treatment of prion diseases. These disorders are invariably fatal. The inherited prion diseases can be prevented by genetic counseling coupled with prenatal DNA screening but such testing may present ethical problems. For example, during childbearing years parents are generally symptom free and may not want to know their own genotype. The apparent incomplete penetrance of some of the inherited prion diseases makes predicting the future for an asymptomatic individual uncertain (GABIZON et al. 1993; GOLDFARB et al. 1991b; HSIAO et al. 1991a).

Ablation of the PrP gene in Tg ($Prnp^{0/0}$) mice has not affected the development of these animals, and they remain healthy at 2 years of age (BÜELER et al. 1992). Since the absence of PrP^{C} expression does not provoke disease, we can conclude that scrapie and other prion diseases are a consequence of PrP^{Sc} accumulation rather than an inhibition of PrP^{C} function. To date the function of PrP^{C} remains unknown.

The resistance of $Prnp^{0/0}$ mice to prions and their failure to propagate scrapie infectivity has prompted suggestion that gene therapy or antisense oligonucleotides might ultimately provide an effective therapeutic approach (BÜELER et al. 1993; PRUSINER et al. 1993b). Mice heterozygous ($Prnp^{0/+}$) for ablation of the PrP gene have prolonged incubation times when inoculated with mouse prions. This finding is in accord with studies on Tg(SHaPrP) mice where decreased SHaPrP expression was accompanied by prolonged incubation times (PRUSINER et al. 1990).

Because the delivery of therapeutic polynucleotides to the CNS remains problematic, the most effective therapy may evolve from the development of drugs which block the conversion of PrP^{C} into PrP^{Sc}. Since the fundamental event in both the formation of PrP^{Sc} and the propagation of prions seems to be the unfolding of α-helices and their refolding into β-sheets (PAN et al. 1993), drugs targeting this structural transformation would seem likely to be efficacious.

Acknowledgments. The work was supported by grants from the National Institutes of Health (NS14069, AG08967, AG02132, NS22786, and AG10770) and the American Health Assistance Foundation, as well as by gifts from the Sherman Fairchild Foundation and the Bernard Osher Foundation.

References

Alper T, Cramp WA, Haig DA, Clarke MC (1967) Does the agent of scrapie replicate without nucleic acid? Nature 214: 764–766

Baker HF, Poulter M, Crow TJ, Frith CD, Lofthouse R, Ridley RM (1991) Amino acid polymorphism in human prion protein and age at death in inherited prion disease. Lancet 337: 1286

Bertoni JM, Brown P, Goldfarb L, Gajdusek D, Omaha NE (1992) Familial Creutzfeldt–Jakob disease with the PRNP codon 200^{lys} mutation and supranuclear palsy but without myoclonus or periodic EEG complexes. Neurology 42 [4, Suppl 3]: 350 (abstract)

Bockman JM, Prusiner SB, Tateishi J, Kingsbury DT (1987) Immunoblotting of Creutzfeldt-Jakob disease prion proteins: host species-specific epitopes. Ann Neurol 21: 589–595

Borchelt DR, Scott M, Taraboulos A, Stahl N, Prusiner SB (1990) Scrapie and cellular prion proteins differ in their kinetics of synthesis and topology in cultured cells. J Cell Biol 110: 743–752

Brown P, Rodgers-Johnson P, Cathala F, Gibbs CJ Jr, Gajdusek DC (1984) Creutzfeldt-Jakob disease of long duration: clinicopathological characteristics, transmissibility, and differential diagnosis. Ann Neurol 16: 295–304

Brown P, Cathala F, Castaigne P, Gajdusek DC (1986) Creutzfeldt-Jakob disease: clinical analysis of a consecutive series of 230 neuropathologically verified cases. Ann Neurol 20: 597–602

Brown P, Goldfarb LG, McCombie WR, Nieto A, Squillacote D, Sheremata W, Little BW, Godec MS, Gibbs CJ Jr, Gajdusek DC (1992) Atypical Creutzfeldt-Jakob disease in an American family with an insert mutation in the PRNP amyloid precursor gene. Neurology 42: 422–427

Brown P, Kaur P, Sulima MP, Goldfarb LG, Gibbs CJJ, Gajdusek DC (1993) Real and imagined clincopathological limits of "prion dementia." Lancet 341: 127–129

Büeler H, Fischer M, Lang Y, Bluethmann H, Lipp H-P, DeArmond SJ, Prusiner SB, Aguet M, Weissmann C (1922) Normal development and behaviour of mice lacking the neuronal cell-surface PrP protein. Nature 356: 577–582

Büeler H, Aguzzi A, Sailer A, Greiner R-A, Autenried P, Aguet M, Weissmann C (1993) Mice devoid of PrP are resistant to scrapie. Cell 73: 1339–1347

Carlson GA, Kingsbury DT, Goodman PA, Coleman S, Marshall ST, DeArmond SJ, Westaway D, Prusiner SB (1986) Linkage of prion protein and scrapie incubation time genes. Cell 46: 503–511

Collinge J, Palmer M (1991) CJD discrepancy. Nature 353: 802

Collinge J, Harding AE, Owen F, Poulter M, Lofthouse R, Boughey AM, Shah T, Crow TJ (1989) Diagnosis of Gerstmann-Sträussler syndrome in familial dementia with prion protein gene analysis. Lancet 2: 15–17

Collinge J, Owen F, Poulter H, Leach M, Crow T, Rosser M, Hardy J, Mullan H, Janota I, Lantos P (1990) Prion dementia without characteristic pathology. Lancet 336: 7–9

Collinge J, Brown J, Hardy J, Mullan M, Rossor MN, Baker H, Crow TJ, Lofthouse R, Poulter M, Ridley R, Owen F, Bennett C, Dunn G, Harding AE, Quinn N, Doshi B, Roberts GW, Honavar M, Janota I, Lantos PL (1992) Inherited prion disease with 144 base pair gene insertion. II. Clinical and pathological features. Brain 115: 687–710

Collinge J, Palmer MS, Campbell T, Sidle KCL, Carroll D, Harding A (1993) Inherited prion disease (PrP lysine 200) in Britain: two case reports. Br Med J 306: 391–392

Come JH, Fraser PE, Lansbury PT Jr (1993) A kinetic model for amyloid formation in the prion diseases: importance of seeding. Proc Natl Acad Sci USA 90: 5959–5963

Cousens SN, Harries-Jones R, Knight R, Will RG, Smith PG, Matthews WB (1990) Geographical distribution of cases of Creutzfeldt-Jakob disease in England and Wales 1970–84. J Neurol Neurosurg Psychiatry 53: 459–465

Crow TJ, Collinge J, Ridley RM, Baker HF, Lofthouse R, Owen F, Harding AE (1990) Mutations in the prion gene in human transmissible dementia. Seminar on molecular approaches to research in spongiform encephalopathies in man, Medical Research Council, London, 14 December 1990 (abstract)

Davison C, Rabiner AM (1940) Spastic pseudosclerosis (disseminated encephalomyelopathy; corticopal-lidospinal degeneration). Familial and nonfamilial incidence (a clinico-pathologic study). Arch Neurol Psychiatry 44: 578–598

Dlouhy SR, Hsiao K, Farlow MR, Foroud T, Conneally PM, Johnson P, Prusiner SB, Hodes ME, Ghetti B (1992) Linkage of the Indiana kindred of Gerstmann-Sträussler-Scheinker disease to the prion protein gene. Nat Genet 1: 64–67

Doh-ura K, Tateishi J, Sasaki H, Kitamoto T, Sakaki Y (1989) Pro → Leu change at position 102 of prion protein is the most common but not the sole mutation related to Gerstmann-Sträussler syndrome. Biochem Biophys Res Commun 163: 974–979
Doh-ura K, Kitamoto T, Sakaki Y, Tateishi J (1991) CJD discrepancy. Nature 353: 801–802
Forloni G, Angeretti N, Chiesa R, Monzani E, Salmona M, Bugiani O, Tagliavini F (1993) Neurotoxicity of a prion protein fragment. Nature 362: 543–546
Friede RL, DeJong RN (1964) Neuronal enzymatic failure in Creutzfeldt-Jakob disease. A familial study. Arch Neurol 10: 181–195
Gabizon R, Meiner Z, Cass C, Kahana E, Kahana I, Avrahami D, Abramsky O, Scarlato G, Prusiner SB, Hsiao KK (1991) Prion protein gene mutation in Libyan Jews with Creutzfeldt-Jakob disease. Neurology 41: 160
Gabizon R, Rosenmann H, Meiner Z, Kahana I, Kahana E, Shugart Y, Ott J, Prusiner SB (1993) Mutation and polymorphism of the prion protein gene in Libyan Jews with Creutzfeldt-Jakob disease. Am J Hum Genet 33: 828–835
Gajdusek DC (1977) Unconventional viruses and the origin and disappearance of kuru. Science 197: 943-960
Gajdusek DC (1985) Subacute spongiform virus encephalopathies caused by unconventional viruses. In: Maramorosch K, McKelvey JJ Jr (eds) Subviral pathogens of plants and animals: viroids and prions. Academic, Orlando, pp 483–544
Gasset M, Baldwin MA, Lloyd D, Gabriel J-M, Holtzman DM, Cohen F, Fletterick R, Prusiner SB (1992) Predicted α-helical regions of the prion protein when synthesized as peptides form amyloid. Proc Natl Acad Sci USA 89: 10940–10944
Gerstmann J, Sträussler E, Scheinker I (1936) über eine eigenartige hereditär-familiäre Erkrankung des Zentralnervensystems Zugleich ein Beitrag zur Frage des vorzeitigen lokalen Alterns. Z Neurol 154: 736–762
Ghetti B, Tagliavini F, Masters CL, Beyreuther K, Giaccone G, Verga L, Farlo MR, Conneally PM, Dlouhy SR, Azzarelli B, Bugiani O (1989) Gerstmann-Sträussler-Scheinker disease. II. Neurofibrillary tangles and plaques with PrP-amyloid coexist in an affected family. Neurology 39: 1453–1461
Giaccone G, Tagliavini F, Verga L, Frangione B, Farlow MR, Bugiani O, Ghetti B (1990) Neurofibrillary tangles of the Indiana kindred of Gerstmann-Sträussler-Scheinker disease share antigenic determinants with those of Alzheimer disease. Brain Res 530: 325–329
Goldfarb L, Brown P, Goldgaber D, Garruto R, Yanaghiara R, Asher D, Gajdusek DC (1990a) Identical mutation in unrelated patients with Creutzfeldt-Jakob disease. Lancet 336: 174–175
Goldfarb LG, Brown P, Goldgaber D, Asher DM, Rubenstein R, Brown WT, Piccardo P, Kascsak RJ, Boellaard JW, Gajdusek DC (1990b) Creutzfeldt-Jakob disease and kuru patients lack a mutation consistently found in the Gerstmann-Sträussler-Scheinker syndrome. Exp Neurol 108: 247–250
Goldfarb LG, Mitrova E, Brown P, Toh BH, Gajdusek DC (1990c) Mutation in codon 200 of scrapie amyloid protein gene in two clusters of Creutzfeldt-Jakob disease in Slovakia. Lancet 336: 514–515
Goldfarb LG, Brown P, McCombie WR, Goldgaber D, Swergold GD, Wills PR, Cervenakova L, Baron H, Gibbs CJJ, Gajdusek DC (1991a) Transmissible familial Creutzfeldt-Jakob disease associated with five, seven, and eight extra octapeptide coding repeats in the PRNP gene. Proc Natl Acad Sci USA 88: 10926-10930
Goldfarb LG, Brown P, Mitrova E, Cervenakova L, Goldin L, Korczyn AD, Chapman J, Galvez S, Cartier L, Rubenstein R, Gajdusek DC (1991b) Creutzfeldt-Jacob disease associated with the PRNP codon 200^{Lys} mutation: an analysis of 45 families. Eur J Epidemiol 7: 477–486
Goldfarb LG, Haltia M, Brown P, Nieto A, Kovanen J, McCombie WR, Trapp S, Gajdusek DC (1991c) New mutation in scrapie amyloid precursor gene (at codon 178) in Finnish Creutzfeldt-Jakob kindred. Lancet 337: 425
Goldfarb LG, Brown P, Haltia M, Cathala F, McCombie WR, Kovanen J, Cervenakova L, Goldin L, Nieto A, Godec MS, Asher DM, Gajdusek DC (1992a) Creutzfeldt-Jakob disease cosegregates with the codon 178^{Asn} PRNP mutation in families of European origin. Ann Neurol 31: 274–281
Goldfarb LG, Petersen RB, Tabaton M, Brown P, LeBlanc AC, Montagna P, Cortelli P, Julien J, Vital C, Pendelbury WW, Haltia M, Wills PR, Hauw JJ, McKeever PE, Monari L, Schrank B, Swergold GD, Autilio-Gambetti L, Gajdusek DC, Lugaresi E, Gambetti P (1992b) Fatal familial insomnia and familial Creutzfeldt-Jakob disease: disease phenotype determined by a DNA polymorphism. Science 258: 806–808
Goldfarb LG, Brown P, Haltia M, Ghiso J, Frangione B, Gajdusek DC (1993) Synthetic peptides corresponding to different mutated regions of the amyloid gene in familial Creutzfeldt-Jakob

disease show enhanced in vitro formation of morphologically different amyloid fibrils. Proc Natl Acad Sci USA 90: 4451–4454

Goldgaber D, Goldfarb LG, Brown P, Asher DM, Brown WT, Lin S, Teener JW, Feinstone SM, Rubenstein R, Kascsak RJ, Boellaard JW, Gajdusek DC (1989) Mutations in familial Creutzfeldt-Jakob disease and Gerstmann-Sträussler-Scheinker's syndrome. Exp Neurol 106: 204–206

Goldhammer Y, Gabizon R, Meiner Z, Sadeh M (1993) An Israeli family with Gerstmann-Sträussler-Scheinker disease manifesting the codon 102 mutation in the prion protein gene. Neurology 43: 2718–2719

Harries-Jones R, Knight R, Will RG, Cousens S, Smith PG, Matthews WB (1988) Creutzfeldt-Jakob disease in England and Wales, 1980–1984: a case-contol study of potential risk factors. J Neurol Neurosurg Psychiatry 51: 1113–1119

Heston LL, Lowther DLW, Leventhal CM (1966) Alzheimer's disease: a family study. Arch Neurol 15: 225–233

Hsiao K, Prusiner SB (1990) Inherited human prion diseases. Neurology 40: 1820–1827

Hsiao K, Baker HF, Crow TJ, Poulter M, Owen F, Terwilliger JD, Westaway D, Ott J, Prusiner SB (1989a) Linkage of a prion protein missense variant to Gerstmann-Sträussler syndrome. Nature 338: 342–345

Hsiao KK, Doh-ura K, Kitamoto T, Tateishi J, Prusiner SB (1989b) A prion protein amino acid substitution in ataxic Gerstmann-Sträussler syndrome. Ann Neurol 26: 137

Hsiao KK, Scott M, Foster D, Groth DF, DeArmond SJ, Prusiner SB (1990) Spontaneous neurodegeneration in transgenic mice with mutant prion protein. Science 250: 1587–1590

Hsiao K, Meiner Z, Kahana E, Cass C, Kahana I, Avrahami D, Scarlato G, Abramsky O, Prusiner SB, Gabizon R (1991a) Mutation of the prion protein in Libyan Jews with Creutzfeldt-Jakob disease. N Engl J Med 324: 1091–1097

Hsiao KK, Cass C, Schellenberg GD, Bird T, Devine-Gage E, Wisniewski H, Prusiner SB (1991b) A prion protein variant in a family with the telencephalic form of Gerstmann-Sträussler-Scheinker syndrome. Neurology 41: 681–684

Hsiao K, Dlouhy S, Farlow MR, Cass C, Da Costa M, Conneally M, Hodes ME, Ghetti B, Prusiner SB (1992) Mutant prion proteins in Gerstmann-Sträussler-Scheinker disease with neurofibrillary tangles. Nat Genet 1: 68–71

Hunter GD (1972) Scrapie: a prototype slow infection. J Infect Dis 125: 427–440

Ikeda S, Yanagisawa N, Allsop D, Glenner GG (1991) A variant of Gerstmann-Sträussler-Scheinker disease with β-protein epitopes and dystrophic neurites in the peripheral regions of PrP – immunoreactive amyloid plaques. In: Natvig JB, Forre O, Husby G, Husebekk A, Skogen B, Sletten K, Westermark P (eds) Amyloid and amyloidosis 1990. Kluwer Academic, Dordrecht, pp 737–740

Jacob H, Pyrkosch W, Strube H (1950) Die erbliche Form der Creutzfeldt-Jakobschen Krankheit. Arch Psychiatr Z Neurol 184: 653–674

Kahana E, Milton A, Braham J, Sofer D (1974) Creutzfeldt-Jakob disease: focus among Libyan Jews in Israel. Science 183: 90–91

Kahana E, Zilber N, Abraham M (1991) Do Creutzfeldt-Jakob disease patients of Jewish Libyan origin have unique clinical features? Neurology 41: 1390–1392

Kirschbaum WR (1924) Zwei eigenartige Erkrankungen des Zentralnervensystems nach Art der spastischen Pseudosklerose (Jakob). Z Ges Neurol Psychiatr 92: 175–220

Kitamoto T (1993) Human prion diseases with PRNP polymorphisms. London 4

Kitamoto T, Tateishi J (1994) Human prion diseases with variant prion protein. Philos Trans R Soc Lond [Biol] 343: 391–398

Kitamoto T, Iizuka R, Tateishi J (1993a) An amber mutation of prion protein in Gerstmann-Sträussler syndrome with mutant PrP plaques. Biochem Biophys Res Commun 192: 525–531

Kitamoto T, Ohta M, Doh-ura K, Hitoshi S, Terao Y, Tateishi J (1993b) Novel missense variants of prion protein in Creutzfeldt-Jakob disease or Gerstmann-Sträussler syndrome. Biochem Biophys Res Commun 191: 709–714

Kretzschmar HA, Stowring LE, Westaway D, Stubblebine WH, Prusiner SB, DeArmond SJ (1986) Molecular cloning of a human prion protein cDNA. DNA 5: 315–324

Kretzschmar HA, Honold G, Seitelberger F, Feucht M, Wessely P, Mehraein P, Budka H (1991) Prion protein mutation in family first reported by Gerstmann, Sträussler, and Scheinker. Lancet 337: 1160

Kretzschmar HA, Kufer P, Riethmüller G, DeArmond SJ, Prusiner SB, Schiffer D (1992) Prion protein mutation at codon 102 in an Italian family with Gerstmann-Sträussler-Scheinker syndrome. Neurology 42: 809–810

Lantos PL, McGill IS, Janota I, Doey LJ, Collinge J, Bruce MT, Whatley SA, Anderton BH, Clinton J, Roberts GW, Rossor MN (1992) Prion protein immunocytochemistry helps to establish the true incidence of prion diseases. Neurosci Lett 147: 67–71

Laplanche J-L, Chatelain J, Launay J-M, Gazengel C, Vidaud M (1990) Deletion in prion protein gene in a Moroccan family. Nucleic Acids Res 18: 6745

Little BW, Brown PW, Rodgers-Johnson P, Perl DP, Gajdusek DC (1986) Familial myoclonic dementia masquerading as Creutzfeldt-Jakob disease. Ann Neurol 20: 231–239

Malmgren R, Kurland L, Mokri B, Kurtzke J (1979) The epidemiology of Creutzfeldt-Jakob disease. In: Prusiner SB, Hadlow WJ (eds) Slow transmissible diseases of the nervous system, vol 1. Academic, New York, pp 93–112

Manetto V, Medori R, Cortelli P, Montagna P, Tinuper P, Baruzzi A, Rancurel G, Hauw J-J, Vanderhaeghen J-J, Mailleux P, Bugiani O, Tagliavini F, Bouras C, Rizszuto N, Lugaresi E, Gambetti P (1992) Fatal familial insomnia: clinical and pathological study of five new cases. Neurology 42: 312–319

Masters CL, Harris JO, Gajdusek DC, Gibbs CJ Jr, Bernouilli C, Asher DM (1978) Creutzfeldt-Jakob disease: patterns of worldwide occurrence and the significance of familal and sporadic clustering. Ann Neurol 5: 177–188

Masters CL, Gajdusek DC, Gibbs CJ Jr, Bernouilli C, Asher DM (1979) Familial Creutzfeldt-Jakob disease and other familial dementias: an inquiry into possible models of virus-induced familial diseases. In: Prusiner SB, Hadlow WJ (eds) Slow transmissible diseases of the nervous system, vol 1. Academic, New York, pp 143–194

Medori R, Montagna P, Tritschler HJ, LeBlanc A, Cortelli P, Tinuper P, Lugaresi E, Gambetti P (1992) Fatal familial insomnia: a second kindred with mutation of prion protein gene at codon 178. Neurology 42: 669–670

Meggendorfer F (1930) Klinische und genealogische Beobachtungen bei einem Fall von spastischer Pseudosklerose Jakobs. Z Ges Neurol Psychiatr 128: 337–341

Meyer RK, McKinley MP, Bowman KA, Braunfeld MB, Barry RA, Prusiner SB (1986) Separation and properties of cellular and scrapie prion proteins. Proc Natl Acad Sci USA 83: 2310–2314

Miyazono M, Kitamoto T, Doh-ura K, Iwaki T, Tateishi J (1992) Creutzfeldt-Jakob disease with codon 129 polymorphism (valine): a comparative study of patients with codon 102 point mutation or without mutations. Acta Neuropathol (Berl) 84: 349–354

Nochlin D, Sumi SM, Bird TD, Snow AD, Leventhal CM, Beyreuther K, Masters CL (1989) Familial dementia with PrP-positive amyloid plaques: a variant of Gerstmann-Sträussler syndrome. Neurology 39: 910–918

Oesch B, Westaway D, Wälchli M, McKinley MP, Kent SBH, Aebersold R, Barry RA, Tempst P, Teplow DB, Hood LE, Prusiner SB, Weissmann C (1985) A cellular gene encodes scrapie PrP 27-30 protein. Cell 40: 735–746

Owen F, Poulter M, Lofthouse R, Collinge J, Crow TJ, Risby D, Baker HF, Ridley RM, Hsiao K, Prusiner SB (1989) Insertion in prion protein gene in familial Creutzfeldt-Jakob disease. Lancet 1: 51–52

Owen F, Poulter M, Collinge J, Crow TJ (1990a) Codon 129 changes in the prion protein gene in Caucasians. Am J Hum Genet 46: 1215–1216

Owen F, Poulter M, Shah T, Collinge J, Lofthouse R, Baker H, Ridley R, McVey J, Crow T (1990b) An in-frame insertion in the prion protein gene in familial Creutzfeldt-Jakob disease. Mol Brain Res 7: 273–276

Owen F, Poulter M, Collinge J, Leach M, Lofthouse R, Crow TJ, Harding AE (1992) A dementing illness associated with a novel insertion in the prion protein gene. Mol Brain Res 13: 155–157

Palmer MS, Dryden AJ, Hughes JT, Collinge J (1991) Homozygous prion protein genotype predisposes to sporadic Creutzfeldt-Jakob disease. Nature 352: 340–342

Palmer MS, Mahal SP, Campbell TA, Hill AF, Sidle KCL, Laplanche J-L, Collinge J (1993) Deletions in the prion protein gene are not associated with CJD. Hum Mol Genet 2: 541–544

Pan K-M, Baldwin M, Nguyen J, Gasset M, Serban A, Groth D, Mehlhorn I, Huang Z, Fletterick RJ, Cohen FE, Prusiner SB (1993) Conversion of α-helices into β-sheets features in the formation of the scrapie prion proteins. Proc Natl Acad Sci USA 90: 10962–10966

Pattison IH (1966) The relative susceptibility of sheep, goats and mice to two types of the goat scrapie agent. Res Vet Sci 7: 207–212

Pearlman RL, Towfighi J, Pezeshkpour GH, Tenser RB, Turel AP (1988) Clinical significance of types of cerebellar amyloid plaques in human spongiform encephalopathies. Neurology 38: 1249-1254

Petersen RB, Goldfarb L, Tabaton M, Brown P, LeBlanc A, Montagna P, Cortelli P, Monari L, Autilio-Gambetti L, Gajdusek DC, Lugaresi E, Gambetti P (1992) Fatal familial insomnia and one subtype of familial Creutzfeld-Jakob disease: effect of a polymorphism on a pathogenic mutation in the prion protein. FASEB J 7: A627

Pocchiari M, Salvatore M, Cutruzzola F, Genuardi M, Travaglini Allocatelli C, Masullo C, Macchi G, Alema G, Galgani S, Xi YG, Petraroli P, Silvestrini MC, Brunori M (1993) A new point mutation of the prion protein gene in familial and sporadic cases of Creutzfeldt-Jakob disease. Ann Neurol 34: 802–807

Poulter M, Baker HF, Frith CD, Leach M, Lofthouse R, Ridley RM, Shah T, Owen F, Collinge J, Brown G, Hardy J, Mullan MJ, Harding AE, Bennett C, Doshi R, Crow TJ (1992) Inherited prion disease with 144 base pair gene insertion. I. Genealogical and molecular studies. Brain 115: 675–685

Prusiner SB (1982) Novel proteinaceous infectious particles cause scrapie Science 216: 136–144

Prusiner SB (1989) Scrapie prions. Annu Rev Microbiol 43: 345–374

Prusiner SB (1991) Molecular biology of prion diseases. Science 252: 1515–1522

Prusiner SB (1993) Genetic and infectious prion diseases. Arch Neurol 50: 1129–1153

Prusiner SB, McKinley MP, Groth DF, Bowman KA, Mock NI, Cochran SP, Masiarz FR (1981) Scrapie agent contains a hydrophobic protein. Proc Natl Acad Sci USA 78: 6675–6679

Prusiner SB, Groth DF, Bolton DC, Kent SB, Hood LE (1984) Purification and structural studies of a major scrapie prion protein. Cell 38: 127–134

Prusiner SB, Scott M, Foster D, Pan K-M, Groth D, Mirenda C, Torchia M, Yang S-L, Serban D, Carlson GA, Hoppe PC, Westaway D, DeArmond SJ (1990) Transgenetic studies implicate interactions between homologous PrP isoforms in scrapie prion replication. Cell 63: 673–686

Prusiner SB, Fuzi M, Scott M, Serban D, Serban H, Taraboulos A, Gabriel J-M, Wells G, Wilesmith J, Bradley R, DeArmond SJ, Kristensson K (1993a) Immunologic and molecular biological studies of prion proteins in bovine spongiform encephalopathy. J Infect Dis 167: 602–613

Prusiner SB, Groth D, Serban A, Koehler R, Foster D, Torchia M, Burton D, Yang S-L, DeArmond SJ (1993b) Ablation of the prion protein (PrP) gene in mice prevents scrapie and facilitates production of anti-PrP antibodies. Proc Natl Acad Sci USA 90: 10608–10612

Puckett C, Concannon P, Casey C, Hood L (1991) Genomic structure of the human prion protein gene. Am J Hum Genet 49: 320–329

Ripoll L, Laplanche J-L, Salzmann M, Jouvet A, Planques B, Dussaucy M, Chatelain J, Beaudry P, Launay J-M (1993) A new point mutation in the prion protein gene at codon 210 in Creutzfeldt-Jakob disease. Neurology 43: 1934–1938

Rosenthal NP, Keesey J, Crandall B, Brown WJ (1976) Familial neurological disease associated with spongiform encephalopathy. Arch Neurol 33: 252–259

Scott M, Foster D, Mirenda C, Serban D, Coufal F, Wälchli M, Torchia M, Groth D, Carlson G, DeArmond SJ, Westaway D, Prusiner SB (1989) Transgenic mice expressing hamster prion protein produce species-specific scrapie infectivity and amyloid plaques. Cell 59: 847–857

Scott M, Groth D, Foster D, Torchia M, Yang S-L, DeArmond SJ, Prusiner SB (1993) Propagation of prions with artificial properties in transgenic mice expressing chimeric PrP genes. Cell 73: 979–988

Sigurdsson B (1954) Rida, a chronic encephalitis of sheep with general remarks on infections which develop slowly and some of their special characteristics. Br Vet J 110: 341–354

Sparkes RS, Simon M, Cohn VH, Fournier REK, Lem J, Klisak I, Heinzmann C, Blatt C, Lucero M, Mohandas T, DeArmond SJ, Westaway D, Prusiner SB, Weiner LP (1986) Assignment of the human and mouse prion protein genes to homologous chromosomes. Proc Natl Acad Sci USA 83: 7358–7362

Stahl N, Baldwin MA, Teplow DB, Hood L, Gibson BW, Burlingame AL, Prusiner SB (1993) Structural analysis of the scrapie prion protein using mass spectrometry and amino acid sequencing. Biochemistry 32: 1991–2002

Stender A (1930) Weitere Beiträge zum Kapitel "Spastische Pseudosklerose Jakobs." Z Neurol Psychiatr 128: 528–543

Tagliavini F, Prelli F, Ghisto J, Bugiani O, Serban D, Prusiner SB, Farlow MR, Ghetti B, Frangione B (1991) Amyloid protein of Gerstmann-Sträussler-Scheinker disease (Indiana kindred) is an 11-kd fragment of prion protein with an N-terminal glycine at codon 58. EMBO J 10: 513–519

Tateishi J, Kitamoto T (1993) Developments in diagnosis for prion diseases. Br Med Bull 49: 971–979

Tateishi J, Kitamoto T, Doh-ura K, Sakaki Y, Steinmetz G, Tranchant C, Warter JM, Heldt N (1990) Immunochemical, molecular genetic, and transmission studies on a case of Gerstmann-Sträussler-Scheinker syndrome. Neurology 40: 1578–1581

Tateishi J, Doh-ura K, Kitamoto T, Tranchant C, Steinmetz G, Warter JM, Boellaard JW (1992) Prion protein gene analysis and transmission studies of Creutzfeldt-Jakob disease. In: Prusiner SB, Colinge J, Powell J, Anderton B (eds) Prion diseases of humans and animals. Horwood, London, pp 129–134

Udovitch AL, Valensi L (1984) The last Arab jews: the communities of Jerba, Tunisia. Harwood Academic, London

Vnencak-Jones CL, Phillips JA (1992) Identification of heterogeneous PrP gene deletions in controls by detection of allele-specific heteroduplexes (DASH). Am J Hum Genet 50: 871–872

Westaway D, DeArmond SJ, Cayetano-Canlas J, Groth D, Foster D, Yang S-L, Torchia M, Carlson GA, Prusiner SB (1994) Degeneration of skeletal muscle, peripheral nerves, and the central nervous system in transgenic mice overexpressing wild-type prion proteins. Cell 76: 117–129

Wilesmith JW, Hoinville LJ, Ryan JBM, Sayers AR (1992) Bovine spongiform encephalopathy: aspects of the clinical picture and analyses of possible changes 1986–1990. Vet Rec 130: 197–201

Fatal Familial Insomnia and Familial Creutzfeldt-Jakob Disease: A Tale of Two Diseases with the Same Genetic Mutation

P. GAMBETTI

"Nick: Fatal familial insomnia .... what sort of disease is it?...
Lili: For a supposedly educated man, you certainly can act ignorant sometimes! Even if you don't know the disease as a clinical entity, you should have read about it ..."

Hilary Koprowski, *"ON NE BADINE PAS AVEC l'ESPRIT CREATIF" or THE SEARCH FOR SCIENTIFIC TRUTH*

1 Fatal Familial Insomnia and Creutzfeldt-Jakob Disease: One Mutation, Two Diseases

Fatal Familial Insomnia (FFI) is a new entry into the group of the prion diseases (MEDORI et al. 1992). FFI is characterized clinically by marked decrease or loss of ability to sleep, dysautonomia and motor signs, and pathologically by preferential atrophy of thalamic nuclei (GAMBETTI et al. 1993). FFI is linked to a mutation a codon 178 of the prion protein gene (*PRNP*) resulting in the substitution of aspartic acid with asparagine (D178N) (MEDORI et al. 1992).

FFI shares the D178N mutation with a subtype of familial Creutzfeldt-Jakob Disease (CJD) referred to as CJD^{178} (GOLDFARB et al. 1992). Despite the common mutation, FFI and CJD^{178} have markedly different clinical and pathological phenotypes (Table 1) and provide an excellent example of phenotypic heterogeneity. Phenotypic heterogeneity is defined as variability in penetrance or expressivity of clinical and pathological features among individuals who share the same genetic

Division of Neuropathology, Institute of Pathology, Case Western Reserve University, 2085 Adelbert Road, Cleveland, OH 44106, USA

Table 1. Phenotype of FFI and CJD[178]

	FFI	CJD[178]
Age at onset	50+9[a] 43±8[b]	39±8[a] 49±4[b]
Duration	11±4[a] 25±7[b]	14±4[a] 26±13[b]
Similar Features		
Cerebellar signs	+	+
Myoclonus	+	+
EEG	Slowing	Slowing
Distinctive Features		
Insomnia	+	–
Dysautonomia	+	–[c]
Preferential thalamic atrophy	+	–
Widespread neocortical spongiosis	±	+
Spongiosis of deep nuclei	–	+

[a] Values for 129 homozygotes.
[b] Values for 129 heterozygotes.
[c] Not reported in clinical histories.

mutation (Steel 1993). It is commonly explained with three mechanisms (Steel 1993). The first, and most important, is genomic imprinting, which is the different expression of a genetic defect depending on whether it is inherited from the mother or the father of the affected subject. Huntington's chorea, spinal cerebellar degeneration, and neurofibromatosis I and II are examples of such a mechanisms (Wexler et al. 1991). In several diseases the genetic defect associated with the genomic imprinting has been shown to be a trinucleotide repeat which is expanded more readily and therefore causes an earlier and more severe disease when it is inherited from one parent rather than the other (McDaniel 1990; Mednick et al. 1994; Medori et al. 1992a). Other explanations of the phenotypic heterogeneity invoke the influence of environmental factors or the effect of modifying genes, i.e., other genes that modify the phenotype expressed by the main mutation (Steel 1993).

To explain the phenotypic heterogeneity between the FFI and CJD[178] we examined the genotypes of 15 patients with the FFI phenotype from five apparently unrelated kindreds and of 15 patients with the CJD[178] phenotype from six apparently unrelated kindreds (Goldfarb et al. 1992). We paid special attention to a common methionine/valine polymorphism at codon 129 of *PRNP* (Fig. 1; Goldfarb et al. 1989, 1992; Collinge et al. 1991). All FFI patients were found to have the methionine codon at position 129 of the *PRNP* mutant allele resulting in the D178N, M/V 129M haplotype (Goldfarb et al. 1992). In contrast, all CJD[178] patients had the valine codon at that position, and the D178N, M/V 129V haplotype (Goldfarb et al. 1992). Both patient populations had either the methionine or the valine codon at position 129 of the normal allele. This resulted in subgroups of 129 homozygous and heterozygous patients in each of the two diseases (Goldfarb et al. 1992). In FFI subjects homozygous at codon 129 the disease had

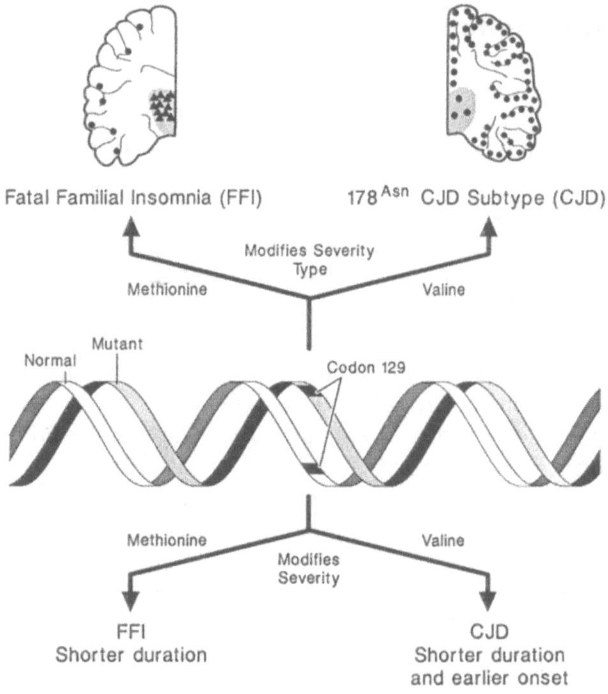

Fig. 1. Diagram of the mechanism of the methionine/valine polymorphism at codon 129 of the prion protein gene (*PRNP*). On the normal allele, the 129 polymorphism determines topography and type of lesions, hence the different phenotype of the disease linked to the mutation at codon 178 of *PRNP*. The presence of the methionine codon results in the expression of the FFI phenotype; the presence of the valine codon in that of CJD^{178}. On the normal allele, the 129 polymorphism influences the duration of FFI and duration and age at onset of CJD^{178}. ●, Spongiform degeneration; Δ, neuronal loss and astrogliosis

a shorter course, while in CJD^{178} homozygotes the disease had not only a shorter course but also an earlier onset. We propose a novel mechanism of phenotypic heterogeneity based on the 129 polymorphism of *PRNP* (Fig. 1). The polymorphism modulates the phenotype of the disease linked to the D178N mutation with the 129 codon located on the mutated as well as with that located on the normal allele. On the mutated allele the 129 polymorphism controls topography and type of the pathology of the D178N-linked prion diseases. In FFI the pathology consists of neuronal loss and gliosis preferentially located in the thalamus; in CJD^{178} the pathology is widespread, involves especially cerebral cortex and basal ganglia, and consists of spongiform degeneration. On the normal allele the 129 polymorphism, by determining age of onset and/or disease duration, controls the severity of the two diseases. All these findings have recently been confirmed in four FFI and one CJD^{178} newly identified kindreds.

2 Phenotypic Heterogeneity: Mechanism

The hallmark of all prion diseases is the presence of an aberrant isoform of a normal glycoprotein called prion protein (PrP) that is partially resistant to proteases (PrP^{res}; Prusiner and DeArmond 1994). It has been suggested that the resistance to proteases is the result of conformational change of the prion protein, and it has also been proposed that PrP that differ in conformation determine phenotypically distinct prion diseases (Prusiner and DeArmond 1994). The mechanism by which the codon 129 on the mutant allele determines the phenotype of FFI and CJD^{178} was further investigated by comparing the PrP^{res} present in brains of FFI and CJD^{178} subjects by gel electrophoresis and immunoblot (Fig. 2; Monari et al. 1994). PrP^{res} migrated as three bands of 29, 27, and 21 kDa in CJD^{178}, and in two bands of 28 and 26 kDa in FFI (Monari et al. 1994). Deglycosylation of the PrP^{res} from both diseases resulted in a single band of 19 kDa in FFI and 21 kDa in CJD^{178}, the same mobility as the fastest moving band of the nondeglycosylated PrP^{res}. Thus, the different PrP^{res} forms result from different degrees of glycosylation of the same PrP backbone. Moreover, the PrP backbone shows a difference in the electrophoretic mobility of approximately 2 kDa between FFI and CJD^{178}. Comparative studies of the individual PrP^{res} fragments generated with endoproteases show that the different mobility of PrP^{res} is due exclusively to the PrP^{res} fragment that contains the N-terminal, which is also the site fo protease cleavage. Therefore, proteinase K, the protease commonly used in these studies, must cleave the PrP^{res} at two different sites in FFI and CJD^{178}, generating PrP^{res} fragments of different sizes in the two diseases (Monari et al. 1994). The mechanism by which PrP^{res} is cleaved differently by the proteinase K in FFI and CJD^{178} has not been established. A likely possibility is that PrP^{res} has a different conformation and presents different cleavage sites in FFI and CJD^{178}. Therefore, the difference in a single amino acid, methionine or valine at position 129 of the PrP^{res}, has a significant effect on ratio of the glycoforms and conformation of the PrP^{res}. Different PrP^{res} forms, in turn are likely to produce the two different pathologies associated with FFI and CJD^{178}, hence the two different phenotypes of these two diseases.

We turned to the polymorphism on the normal allele to gain insight into the possible mechanisms by which this polymorphism regulates the duration of the disease in FFI. Examination of the clinical phenotype of FFI subjects homozygous

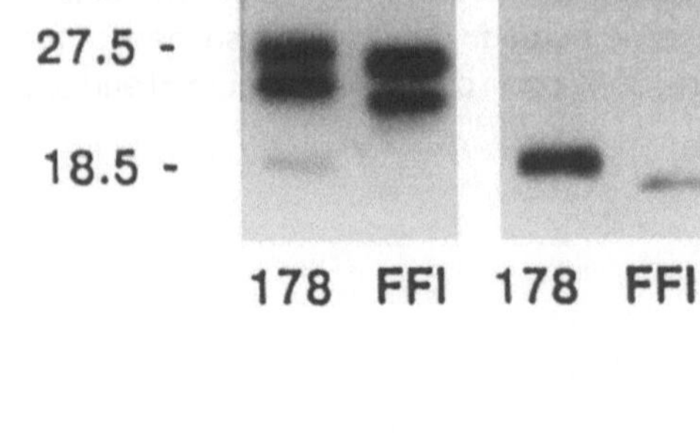

Fig. 2. Western blot of proteinase K resistant prion protein in the subtype of familial CJD with mutation at codon 178 of the prion protein gene (178) and in FFI. Without (–) or with (+) deglycosylation (PNGase). *Left*, three bands are detectable in CJD^{178} but only two in FFI; right, deglycosylation shows distinctly that the two proteinase K resistant fragments of the prion protein are of different sizes

and heterozygous at codon 129 reveals that in addition to a different disease duration (Table 1), the two patient populations also have a slightly different clinical presentation and pathology (MANETTO et al. 1992). Clinically, homozygous FFI patients present with insomnia and dysautonomia while motor signs and myoclonus appear later (MANETTO et al. 1992). The opposite is true for the heterozygote patients, who present with motor signs, while insomnia and dysautonomia follow at a later stage (MANETTO et al. 1992). Pathologically, while the thalamic atrophy is similar in both patient populations, spongiform degeneration of the cerebral neocortex is more prominent in the heterozygous than the homozygous subjects (PARCHI et al. 1995). The amount of PrP^{res} was found to be higher in 129 heterozygous subjects than in homozygous subjects in several brain regions such as the cerebral neocortex and the limbic system. This difference is likely to be due to the broader diffusion of the PrP^{res} in 129 heterozygotes due to the longer disease duration in these subjects. This conclusion is supported by the data obtained by positron emission tomography (PET) which show that in the early stages of the disease the hypometabolism is present predominantly in the thalamus in both patient populations, and then spreads progressively to the cortex according to the duration of the disease (PERANI et al. 1993; PERANI et al., unpublished data). However, in contrast to the neocortex, the amount of PrP^{res} present in the brainstem is similar in the homozygous and heterozygous patient populations, despite the different duration of the disease. Thus, the rate or the timing of conversion of soluble PrP into PrP^{res} in the brain stem is different in heterozygotes and homozygotes and it is slower, or later, in FFI heterozygous patients. If FFI patients expire when a critical amount of PrP^{res} is reached in the brainstem, then the codon 129 on the normal allele may regulate the disease duration in FFI by determining the rate or timing of PrP^{res} accumulation in this critical region of the brain.

3 Conclusions

FFI and CJD^{178}, two phenotypically distinct diseases linked to the same D178N mutation in PRNP, provide a novel mechanism of phenotypic heterogeneity determined by the methionine/valine polymorphism at codon 129 of PRNP. On the mutated allele, codon 129 determines the type of disease. FFI is linked to the D178N, M/V M haplotype, while the haplotype of CJD^{178} is D178N M/V 129 V. Moreover, both diseases have a more rapid course in affected subjects homozygous at codon 129. Thus, on the normal allele codon 129 modulates the duration in both FFI and CJD^{178}.

The presence of methionine in FFI and of valine in CJD^{178} at position 129 of the PrP in conjunction with the presence of the mutant asparagine at position 178, results in the expression of two PrP^{res} forms that are distinct for the different ratio of the three glycoforms and the different size of the protein core, probably reflecting a different conformation. The two PrP^{res} forms cause two different pathologies, hence two different phenotypes of FFI and CJD^{178}. Codon 129 on the normal PRNP allele is likely to shorten the disease duration in FFI homozygous subjects by affecting the rate of PrP^{res} accumulation in some brain regions, selectively. The more rapid accumulation of PrP^{res} in the brainstem of homozygotes may be the cause of the shorter disease duration in these patients. On the other hand, the longest disease duration in heterozygote is likely to lead to a more widespread distribution of PrP^{res} in these subjects. Thus, the 129 PRNP codon has a twofold effect on the disease phenotype according to its location in the mutant or in the normal allele. These findings emphasize the control that the genotype may have on the phenotypic expression of genetic diseases and the complexity of this control. Apolipoproteins are another genetic factor that may influence phenotypic expression of prion diseases (Amouyel et al. 1994). A genetic mechanism by which a nonpathogenic polymorphism regulates the phenotype of a disease by influencing cotranslational or posttranslational events of the pathogenic protein such as glycosylation and conformation applies exquisitely to the prion diseases in which the critical role of conformation is supported by strong evidence (Steel 1993). However, it seems likely that a similar mechanism plays a role in other familial degenerative diseases (Steel 1993). Because of their particular features, prion diseases may reveal mechanisms that are operational, but more difficult to detect, in other diseases.

Acknowledgments. This research was supported by grants AG08012, AGNS08155, AG08992, and the Britton Fund.

References

Amouyel P, Vidal U, Launay JM, Laplanche J (1994) The apolipoportein E alleles as major susceptibility factor for Creutzfeldt-Jakob disease. Lancet 344: 1315–1318

Collinge J, Palmer MS, Dryden AJ (1991) Genetic predisposition to iatrogenic Creutzfeldt-Jakob diseae. Lancet 337: 1441–1442

Gambetti P, Petersen R, Monari L, Tabaton M, Cortelli P, Montagna P, Lugaresi E, Autilio-Gambetti L (1993) Fatal familial insomnia and the widening spectrum of prion diseases. In: Allen I (ed) Spongiform Encephalopathies Br Med Bull 49(4): 980–994

Gambetti P, Petersen R, Tabaton M, Gallasi R, Lugaresi E (1995) Thalamic dementia: relationship with fatal familial insomnia. In: The primary degenerative dementias other than Alzheimer's disease

Goldfarb LG, Brown P, Goldgaber D (1989) Patients with Creutzfeldt-Jakob disease and kuru lack the mutation in the PRIP gene found in Gerstmann-Sträussler-Scheinker syndrome, but they show a different double-allele mutation in the same gene. Am J Hum Genect 45 [Suppl]: A189

Goldfarb LG, Petersen RB, Tabaton M, Brown P, LeBlanc AC, Montagna P, Cortelli P, Julien J, Vital C, Pendlebury WW, Haltia M, Willis PR, Hauw JJ, McKeever PE, Monari L, Schrank B, Swergold GD, Autilio-Gambetti L, Gajdusek C, Lugaresi E, Gambetti P (1992) Fatal familial insomnia and Familial Creutzfeldt Jakob disease: disease phenotype determined by a DNA polymorphism. Science 258: 806–808

Manetto V, Medori R, Cortelli P, Montagna P, Baruzzi A, Hauw J, Rancruel G, Vanderhaeghen JJ, Mailleux P, Bugiani O, Tagliavini F, Bouras C, Rizzuto N, Lugaresi E, Gambetti P (1992) Fatal familial insomnia: clinical and pathological study of five new cases. Neurology 2: 312–319

McDaniel KD (1990) Thalamic degeneration. In: Cummings JL (ed) Subcortical Dementia. Oxford University Press, New York, pp 132–144

Mednick AS, Reder AT, Spire JP, Van Cauter E, Brown P, Wollmann RL, Goldfarb LG, Garay A, Ovsiew F, Gajdusek DC, Roos RP (1994) Fatal familial insomnia. Neurology 44 [Suppl 2]: 285

Medori R, Montagna P, Tritschler JH, LeBlanc A, Cortelli P, Lugaresi E, Gambetti P (1992a) Fatal familial insomnia: a second kindred with mutation of prion gene at codon 178. Neurology 42: 669–670

Medori R, Tritschler HJ, LeBlanc A, Villare F, Manetto V, Chen HY, Xue R, Leal S, Montagna P, Cortelli P, Tinuper P, Avoni P, Mochi M, Baruzzi A, Hauw JJ, Ott J, Lugaresi E, Autilio-Gambetti L, Gambetti P (1992b) Fatal familial insomnia is a prion disease with a mutation at codon 178 of the prion gene. N Engl J Med 326: 444–449

Monari L, Chen SC, Brown P, Parchi P, Petersen RB, Mikol J, Gray F, Cortelli P, Montagna P, Ghetti B, Goldfarb LG, Gajdusek DC, Lugaresi E, Gambetti P, Autilio-Gambetti L (1994) Fatal familial insomnia and familial Creutzfeldt-Jakob disease: Different prion proteins determined by a DNA polymorphism. Proc Natl Acad Sci USA 91: 2839–2842

Parchi P, Castellani R, Cortelli P, Montagna P, Chen SG, Petersen RB, Lugaresi E, Autilio-Gambetti L, Gambetti P (1995) Regional distribution of protease resistant prion protein in Fatal Familial Insomnia: correlation with histopathology and polymorphism at codon 129 of the prion protein gene. Ann Neurol 38: 21–29

Perani D, Cortelli P, Lucignani G, Montagna P, Tinuper P, Gallasi R, Gambetti P, Lenzi GL, Lugaresi E, Fazio F (1993) [^{18}F] FDG PET in fatal familial insomnia: the functional effects of thalamic lesions. Neurology 43: 2565–2569

Prusiner SB, DeArmond SJ (1994) Prion diseases and neurodegeneration. Annu Rev Neurosci 17: 311–319

Steel M (1993) Genetics: polymorphism, proteins and phenotypes. Lancet 341: 212–213

Wexler NS, Rose EA, Housman DE (1991) Molecular approaches to hereditary diseases of the nervous system: Huntington's disease as a paradigm. Annu Rev Neurosci 14: 503–529

Human Prion Disease and Human Prion Protein Disease

T. KITAMOTO[1] and J. TATEISHI[2]

1 Introduction

Creutzfeldt-Jakob disease (CJD), kuru, and Gerstmann-Sträussler syndrome (GSS) show clinical and pathological characteristics similar to those of scrapie, a transmissible neurodegenerative disease of sheep and goats. These diseases are caused by slow infectious agents designated as prions (PRUSINER 1982). The major component of prions is prion protein (PrP; MCKINLEY et al. 1983), which is encoded in normal human genomes located on the short arm of chromosome 20 (SPARKES et al. 1986). In 1989 codon 102 or codon 117 point mutations of human PrP were reported to be linked to GSS (HSIAO et al. 1989; DOH-URA et al. 1989). The results in codon 102 transgenic mice also strengthen the idea that this mutation is one of the essential events that cause GSS (HSIAO et al. 1990). The several polymorphisms or mutations were also reported in familial CJD and familial dementia (GOLDGARBER et al. 1989; GOLDFARB et al. 1991; MEDORI et al. 1992; KITAMOTO et al. 1993a,b).

Recently, several studies showed that scrapie form of PrP (PrP^{Sc}) is an essential component of prions. These studies included the following results: copurification of PrP^{Sc} and scrapie infectivity (MCKINLEY et al. 1983; GABIZON et al. 1988), PrP^{Sc} detection only in clones of cultured cell producing prion infectivity (TARABOULOS et al. 1990), PrP amyloid plaque detections in prion diseases (BENDHEIM et al. 1984; KITAMOTO et al. 1986), genetic linkage between human PrP gene mutation and hereditary CJD or GSS (HSIAO et al. 1989, 1992; TRANCHANT et al.

[1] Department of Neurological Science, Tohoku, University School of Medicine, Sendai 980, Japan
[2] Department of Neuropathology, Neurological Institute, Kyushu University 60, Fukuoka 812, Japan

1992; GOLDFARB et al. 1992), and genetic linkage between mouse PrP gene and scrapie incubation time (WESTAWAY et al. 1987). One of the major confirmations is the result in PrP knockout ($PrP^{\%}$) mice (BÜELER et al. 1992). These $PrP^{\%}$ mice did not develop the scrapie and did not amplify the infectivity (prions titer; BÜELER et al. 1993). Therefore, PrP is an essential component for the infectivity. However, the mechanism of prion multiplication is still unclear. One fascinating hypothesis (the prion dimer hypothesis) was proposed by PRUSINER (1991). This hypothesis can explain the infectious form of prion disease. We report here whether the dimer hypothesis can explain the human prion disease with germ-line mutation. We also report a new variant GSS with codon 102 Leu mutation and codon 219 Lys polymorphism.

2 Prion Dimer Hypothesis and Mutant PrP Molecule

The mechanism by which prions multiply is unknown. The multiplication of prion infectivity is an exponential process in which the posttranslational conversion of PrP^{C} (normal cellular form) to PrP^{Sc} appears necessary. According to the prion dimer hypothesis, a PrP^{Sc} molecule combines with one PrP^{C} molecule giving rise to one heterodimer. This heterodimer is subsequently transformed into one homodimer (PrP^{Sc}/PrP^{Sc}) that dissociates to combine with two PrP^{C} molecules creating an exponential process.

In humans with PrP point mutation, mutant PrP^{C} molecules might spontaneously convert into mutant PrP^{Sc} (Fig. 1). While the initial stochastic event may be inefficient, once it happens, the process becomes autocatalytic. Whether all GSS and familial CJD patients contain infectious prions is unknown. If the former is found, mutant PrP^{Sc} molecules combine with the heterodimer (mutant PrP^{Sc}/wild PrP^{C}) and are subsequently transformed into mutant PrP^{Sc}/wild PrP^{Sc}. This wild PrP^{Sc} produces the heterodimer (wild PrP^{Sc}/wild PrP^{C}) in a exponential process (Fig. 1). If the latter is found, presumably, mutant PrP^{Sc} molecules alone can produce the central nervous system dysfunction (Fig. 2). To test the dimer hypothesis we examined the following cases with unique point mutation or polymorphism.

3 Only Mutant PrP^{Sc} Accumulates in the Central Nervous System (Prion Protein Disease)

Most of the point mutations on the PrP gene were heterozygous and missense. Therefore it is difficult to analyze which molecule, wild or mutant, accumulated in the central nervous system because of the identical molecular weights. Previously we and another group identified the mutant PrP molecule (codon 102 Leu

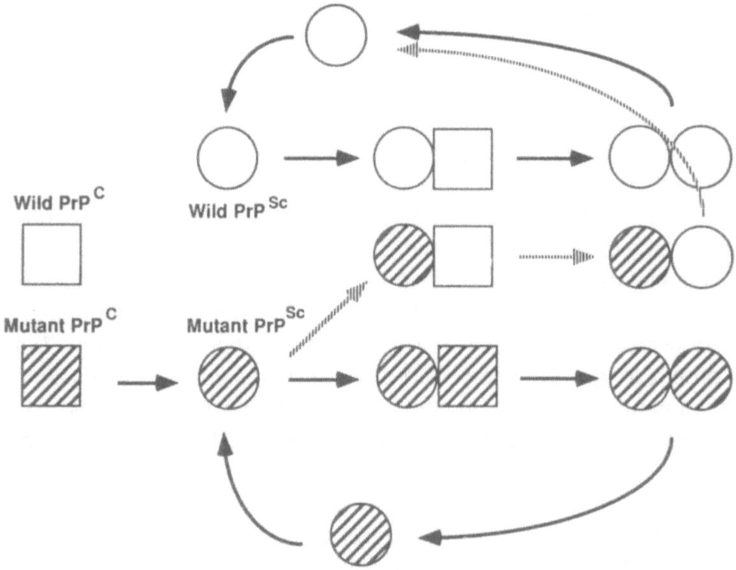

Fig. 1. Prion dimer hypothesis in inherited prion disease in humans. Mutant PrP^C molecule (*dashed line in squares*) might initiate the conversion of PrP^C to PrP^{Sc} (*dashed line in circles*). When infectious prions are produced, they stimulate the synthesis of mutant PrP^{Sc} (*dashed line*) and wild PrP^{Sc} (*open circle*)

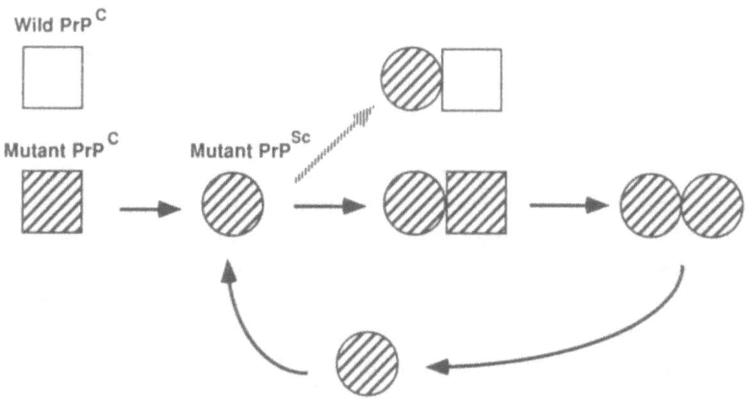

Fig. 2. Inborn error of PrP metabolism in humans. Mutant PrP^{Sc} molecule combine with only mutant PrP^C molecule. Prion infectivity is not generated, but humans develops neurological dysfunction and PrP amyloid plaques

or codon 129 Val/198 Ser) from the kuru plaque core fractions of GSS patients (KITAMOTO et al. 1991; TAGLIAVINI et al. 1991). However, it remains to be established whether mutant PrP are major component of PrP^{Sc}. These peptide-sequencing data were based on a purification step which might cause the fragmentation of

PrP or make it difficult to measure the concentration of mutant PrP in the kuru plaques.

Among these limitations of analyzing the mutant PrP molecules, we had a chance to examine a patient with Y145 stop mutation (Kitamoto et al. 1993a; Fig. 3). The T to G transition at codon 145 was resulted in tyrosine (TAT) to amber codon (TAG). To analyze this stop codon we used the following steps: (a) mRNA expression of mutant PrP in the brain, (b) western blotting, (c) protein expression in *Escherichia coli* and (d) immunohistochemistry using N-terminal and C-terminal antibodies.

First, we checked the mRNA expression with reverse-transcription polymerase chain reaction (RT-PCR) amplification. The RT-PCR product was digested with *Mae*I endonuclease. The mutant PrP gene has a cutting site of *Mae*I. *Mae*I restriction fragment length polymorphism analysis revealed the expression of both wild and mutant PrP mRNA in the brain. Next, we performed western blotting using a proteinase-resistant prion rod fraction in the brain tissue to check for the low molecular weight mutant PrP. The western blot showed a diffuse smear immunoreactivity from the top of the gel to the dye front. The smear immunoreactivity, which may be due to the highly aggregated PrP of kuru plaques, makes it difficult to reveal small molecular weight mutant PrP. Thus we prepared a fusion protein expression plasmid in *E.coli* to check the specificity of the N-terminal and C-terminal PrP antibodies. N-terminal antibody recognized both the wild and mutant PrP fusion protein, and C-terminal antibody recognized only the wild PrP fusion protein. Finally, we examined the mutant PrP molecules in the tissue sections with these PrP antibodies. The N-terminal antibody immunostained positively with kuru plaques in the brain from this patient with Y145 stop, but the C-terminal antibody did not immunolabel the kuru plaque. To confirm the immunoreactivity of the C-terminal antibody, we also immunostained the kuru plaques in the GSS patients with P102L mutation. This C-terminal antibody positively recognized kuru plaques in GSS102. Therefore, in this peculiar case, only mutant PrP molecules aggregated to form kuru plaques.

In this case, mutant PrP^{C} molecules might spontaneously convert into PrP^{Sc}. While the initial stochastic event may be inefficient, once it happens, the process becomes autocatalytic. Mutant PrP molecules alone produced the amyloid plaques in the central nervous system. Therefore this point mutation represents inborn errors of PrP metabolism (prion protein disease). Transmission study from this case to mouse also showed a negative result (0/10 mice).

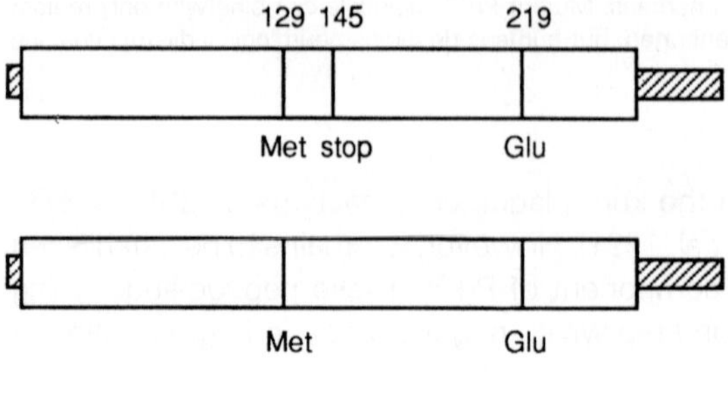

Fig. 3. PrP genotype of a patient with codon 145 mutation. Codon 129 and codon 219 representnormal polymorphism seen in Japanese

4 Mutant PrP^{Sc} Influences the Wild PrP^{Sc} Conversion (Prion Disease)

We have another case with codon 180 Ile mutation and codon 129 Val polymorphism each on the different allele (Fig. 4). In our experience with codon 180 Ile mutation, patients had only Met/Met type polymorphism at codon 129 and Glu/Glu type polymorphism at codon 219. Neuropathological examinations showed typical spongiform changes and moderate neuronal loss in the cerebral cortices. Western blot analysis revealed PrP^{Sc}, but the concentration of PrP^{Sc} was much less than that of the wild-type CJD patients. The wild-type CJD patients have three major PrP^{Sc} bands, but CJD patients with codon 180 Ile have two major bands corresponding to nonglycosylated and one glycosylated PrP^{Sc} (HITOSHI et al. 1993). PrP immunostainings showed diffuse gray matter stainings but not amlyoid plaques. In CJD cases with Val/Met or Val/Val polymorphism at codon 129 we determined the amyloid plaque formation in the central nervous system (KITAMOTO et al. 1992; MIYAZONO et al. 1992). Therefore this patient is a suitable case to examine the both mutant (codon 180 Ile) and wild (codon 129 Val) PrP molecules. Figure 5 shows the working hypothesis for detection of mutant and wild PrP molecules. Mutant PrP^{C} (180 Ile) might convert to mutant PrP^{Sc}. If the conversion occurs, the mutant PrP^{C}/mutant PrP^{Sc} heterodimer formation may result in the two mutant PrP^{Sc} molecules. The interesting point is whether the wild PrP^{C}/mutant PrP^{Sc} heterodimer formation occurs. If the wild PrP^{C} is converted to wild PrP^{Sc}, the process of wild PrP^{Sc} formation becomes autocatalytic. These wild PrP^{Sc} (codon 129 Val) accumulations could be observed in the amyloid plaque formation in the central nervous system.

To examine this working hypothesis we at first analyzed the neuropathological findings of this case. Routine histopathological findings were severe spongiform changes and neuronal loss in the cerebral cortices, but no congophilic amyloid plaques. Western blot showed two lower PrP^{Sc} bands. The results of these histopathological and western blot analyses of this case are compatible with the findings seen in other CJD patients with codon 180 Ile. Finally, we examined PrP immunostainings using the hydrolytic autoclaving pretreatment. PrP immunostainings revealed weakly positive diffuse gray matter stainings and a few amyloid plaques. These amyloid plaques were not observed in other CJD

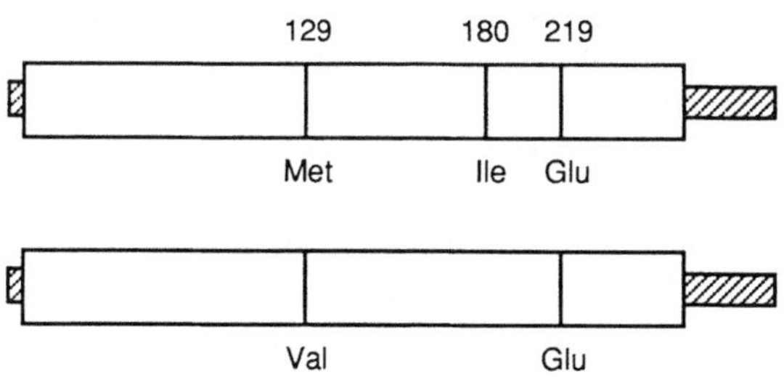

Fig. 4. PrP genotype of a patient with codon 180 mutation and codon 129 Val polymorphism

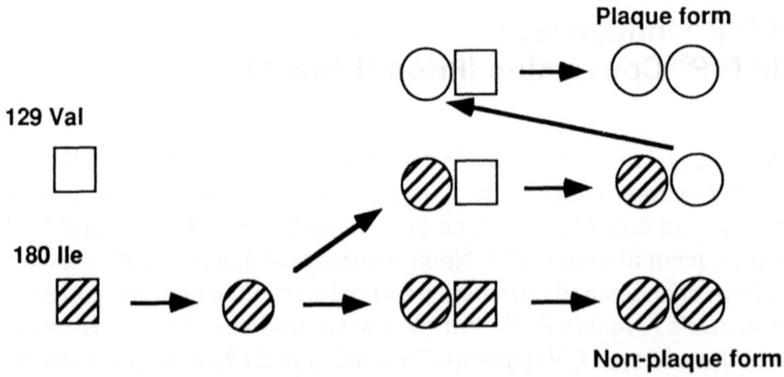

Fig. 5. Prion dimer hypothesis in a patient with codon 180 mutation and codon 129 Val polymorphism each on the different PrP allele

patients with codon 180 Ile mutation and codon 129 Met/Met polymorphism. Therefore in this case wild PrP^{Sc} molecules also accumulated in the central nervous system. Transmission experiment is now continuing.

5 A New Variant PrP Molecule in Gerstmann-Sträussler Syndrome

We determined the new polymorphism (codon 219 Glu or Lys) in a normal Japanese population (KITAMOTO and TATEISHI 1994). The allele frequency of codon 219 Lys is about 6% in normal Japanese. During the search for this polymorphism codon 219 Lys was detected in four patients with Japanese Gerstmann-Sträussler syndrome (P102L). In three patients belonged to the different families, codon 102 Leu mutation was on the codon 219 Glu allele, but not on the codon 219 Lys allele. These three patients have a typical clinical course showing spinocerebellar degeneration. However, in one patient codon 102 Leu mutation was detected on the codon 219 Lys allele (Fig. 6). Family study revealed that four patients in this family have codon 102 Leu and codon 219 Lys on the same allele. Two patients showed only dementia in the absence of cerebellar signs, and two others showed weak cerebellar signs and dementia.

Recently we examined one autopsy case with a clinical course of dementia in the absence of cerebellar signs. Histopathological examinations showed no spongiform changes in the cerebral cortices. Mild gliosis and mild neuronal loss were observed in the deep layer of the cerebral cortices. There were no congophilic amyloid plaques in the cerebral and cerebellar cortices. The absence of amyloid plaques is quite different from GSS patients with only codon 102 Leu. Previously we examined more than 20 GSS patients with codon 102 Leu. These GSS patients have many congophilic plaques in the cerebral and cerebellar

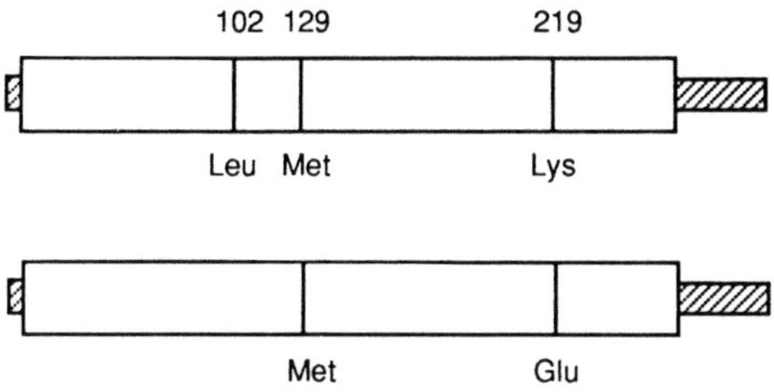

Fig. 6. PrP genotype of a new GSS family in Japan. The codon 102 Leu mutation is located on the codon 219 Lys allele

cortices, but this patient does not have congophilic plaques in routine histopathological examinations. PrP immunostainings revealed a few amyloid plaques in the molecular layer of the cerebellar cortices, and diffuse amyloid plaques, so-called moth-eaten PrP plaques, in the deep cortical layers of the cerebral cortices and basal ganglia. There were neither tau-positive neurofibrillary tangles nor senile plaques in the cerebral cortices.

These clinical and neuropathological findings support the hypothesis that codon 219 Lys polymorphism influences the phenotype of the codon 102 Leu mutation. Therefore it is better to classify as a new type GSS with codon 102 Leu/codon 219 Lys.

Acknowledgments. We thank M. Yoneda and K. Hatanaka for technical assistance. This study was supported by a grant from the Science and Technology Agency, Grant-in-Aid for Scientific Research and a Grant-in-Aid for Scientific Research on Priority Area from the Ministry of Education, Science, and Culture, a grant from the Ministry of Health and Welfare, a grant from the Kato Memorial Trust for Manbyo Research, and a grant from Nambyo Medical Research.

References

Bendheim PE, Barry RA, DeArmond SJ, Stites DP, Prusiner SB (1984) Antibodies to a scrapie prion protein. Nature 310: 418–421

Büeler H, Fischer M, Lang Y, Bluethmann H, Lipp HP, DeArmond SJ, Prusiner SB, Aguet M, Weissmann C (1992) Normal development and behaviour of mice lacking the neuronal cell-surface PrP protein. Nature 356: 577–582

Büeler H, Aguzzi A, Sailer A et al. (1993) Mice devoid of PrP are resistant to scrapie. Cell 73: 1339–1347

Doh-ura K, Tateishi J, Sasaki H, Kitamoto T, Sakaki Y (1989) Pro–Leu change at position 102 of prion protein is the most common but not the sole mutation related to Gerstmann-Sträussler syndrome. Biochem Biophys Res Commun 163: 974–979

Gabizon R, McKinley MP, Groth DF, Prusiner SB (1988) Immunoaffinity purification and neutralization of scrapie prion infectiviy. Proc Natl Acad Sci USA 85: 6617–6621

Goldfarb LG, Haltia M, Brown P, Nieto A, Kovanen J, McCombie WR, Trapp S, Gajdusek DC (1991) New mutation in scrapie amyloid precursor gene (at codon 178) in Finnish Creutzfeldt-Jakob kindred. Lancet 337: 425

Goldfarb LG, Petersen RB, Tabaton M, Brown P, LeBlanc AC, Montagna P, Cortelli P, Julien J, Vital C, Rendelbury WW, Haltia M, Wills PR, Hauw JJ, McKeever PE, Monari L, Schrank B, Swergold GD, Autilio-Gambetti L (1992) Fatal familial insommia and familial Creutzfeldt-Jakob disease: disease phenotype determined by a DNA polymorphism. Science 258: 806–809

Goldgarber D, Goldfarb LG, Brown P et al. (1989) Mutations in familial Creutzfeldt-Jakob disease and Gerstmann-Sträussler syndrome. Exp Neurol 106: 204–206

Hitoshi S, Nagura H, Yamanouchi H, Sakuta M, Kitamoto T (1993) Double mutations at codon 180 and codon 232 of the PRNP gene in an apparently sporadic case of Creutzfeldt-Jakob disease. J Neurol Sci 120: 208–212

Hsiao K, Baker HF, Crow TJ, Poutler M, Owen F, Terwillinger JD, Westaway D, Ott J, Prusiner SB (1989) Linkage of a prion protein missense variant to Gerstmann-Sträussler syndrome. Nature 338: 342–345

Hsiao K, Dlouhy SR, Farlow MR, Cass C, Costa MD, Conneally PM, Hodes ME, Ghetti B, Prusiner SB (1992) Mutant prion proteins in Gerstmann-Sträussler-Scheinker disease with neurofibrillary tangles. Nature Genet 1: 68–71

Hsiao KK, Scott M, Foster D, Groth DF, DeArmond SJ, Prusiner SB (1990) Spontaneous neurodegeneration in transgenic mice with mutant prion protein of Gerstmann-Sträussler syndrome. Science 250: 1587–1590

Kitamoto T, Tateishi J (1994) Human prion diseases with variant prion protein. Philos Trans R Soc Lond [B] 343: 391–398

Kitamoto T, Tateishi J, Tashima T, Takeshita I, Barry RA, DeArmond SJ, Prusiner SB (1986) Amyloid plaques in Creutzfeldt-Jakob disease stain with prion protein antibodies. Ann Neurol 20: 204–208

Kitamoto T, Yamaguchi K, Doh-ura K, Tateishi J (1991) A prion protein missense variant is integrated in kuru plaque cores in patients with Gerstmann-Sträussler syndrome. Neurology 41: 306–310

Kitamoto T, Doh-ura K, Mutamoto T, Miyazono M, Tateishi J (1992) The primary structure of the prion protein influences the distribution of abnormal prion protein in the central nervous system. Am J Pathol 141: 271–277

Kitamoto T, Lizuka R, Tateishi J (1993a) An amber mutation of prion protein in Gerstmann-Sträussler syndrome with mutant PrP plaques. Biochem Biophys Res Commun 192: 525–531

Kitamoto T, Ohta M, Doh-ura K, Hitoshi S, Terao Y, Tateishi J (1993b) Novel missense variants of prion protein in Creutzfeldt-Jakob disease or Gerstmann-Sträussler syndrom. Biochem Biophys Res Commun 191: 709–714

McKinley MP, Boltn DC, Prusiner SB (1983) A protease-resistant protein is a structural component of the scrapie prion. Cell 35: 57–62

Medori R, Tritschler HJ, LeBlanc A, Villare F, Manetto V, Chen HY, Xue R, Leai S, Montagna P, Gortelli P, Tinuper P, Avoni P, Mochi M, Baruzzi A, Hauw JJ, Ott J, Lugaressi E, Autilio-Gambetti L, Gambetti P (1992) Fatal familial insomnia, a prion disease with a mutation at codon 178 of the prion protein gene. N Engl J Med 326: 444–449

Miyazono M, Kitamoto T, Doh-ura K, Iwaki T, Tateishi J (1992) Creutzfeldt-Jakob disease with codon 129 polymorphism (valine): a comparative study of patients with codon 102 point mutation or without mutations. Acta Neuropathol (Berl) 84: 349–354

Prusiner SB (1982) Novel proteinaceous infectious particles cause scrapie. Science 216: 136–144

Prusiner SB (1991) Molecular biology of prion diseases. Science 252: 1515–1522

Sparkes RS, Simon M, Cohn VH, Fournier REK, Lem J, Klisak I, Heinzman C, Blatt C, Lucero M, Mohandas T, DeArmond SJ, Westaway D, Prusiner SB, Weiner LP (1986) Assignment of the human and mouse prion protein genes to homologous chromosomes. Proc Natl Acad Sci USA 83: 7358–7362

Tagliavini F, Prelli F, Ghiso J, Bugiani O, Serban D, Prusiner SB, Farlow MR, Ghetti B, Frangione B (1991) Amyloid protein of Gerstmann-Sträussler-Scheinker disease (Indiana kindred) is an 11kd fragment of prion protein with an N-terminal glycine at codon 58. EMBO J 10: 513–519

Taraboulos A, Serban D, Prusiner SB (1990) Scrapie prion proteins accumulate in the cytoplasm of persistently infected cultured cells. J Cell Biol 110: 2117–2132

Tranchant C, Doh-ura K, Warter JM, Steinmetz G, Chevalier Y, Hanauer A, Kitamoto T, Tateishi J (1992) Gerstmann-Sträussler-Scheinker disease in an Alsatian family: clinical and genetic studies. J Neurol Neurosurg Psychiatry 55: 185–187

Westaway D, Goodman PA, Mirenda CA, McKinley MP, Carlson GA, Prusiner SB (1987) Distinct prion proteins in short and long scrapie incubation period mice. Cell 51: 651–662

Prion Strains

G.A. CARLSON

1 Introduction

More than 30 years ago PATTISON and MILLSON (1961) reported the occurrence of two distinct constellations of clinical signs, labeled nervous or drowsy, in goats experimentally infected with scrapie. These syndromes were reproducible through subsequent intracerebral inoculations, suggesting the existence of "true breeding" variants of scrapie agent. Subsequent transmissions of both goat and sheep scrapie into experimental rodents set the stage for isolation of multiple scrapie "strains" (ZLOTNIK 1965). The existence of scrapie isolates that exhibit distinct and stable properties in the same inbred mouse strain provided evidence for agent-specified information, thought to be encoded by a polynucleotide (BRUCE and DICKINSON 1987). The existence of microbiological strains of agent continues to be presented as an argument against the concept of a proteinaceous infectious particle devoid of functional nucleic acid. The only known functional component of prions is PrP^{Sc}, a disease-specific, posttranslational derivative of the normal prion protein isoform, PrP^C (PRUSINER 1991). β-pleated sheet is a major secondary structure of PrP^{Sc}, in contrast to the predominance of α-helix in PrP^C, and the two isoforms may differ only in their conformations (PAN et al. 1993; SAFAR et al. 1993). PrP^{Sc} is certainly necessary, and possibly sufficient, for transmission of scrapie. No

McLaughlin Research Institute, 1520 23rd Street South, Great Falls MT 59405, USA

scrapie-specific nucleic acid has ever been detected in purified prion preparations, and to be present at one molecule per infectious unit a prion-specific nucleic acid would need to have fewer than 50 nucleotides (KELLINGS et al. 1992). Properties of viral, bacterial and fungal strains are most often attributed directly to mutations within their nucleic acid genomes, with the interesting exception of prion-like epigenetic phenomena in yeast (WICKNER 1994). If prions are devoid of functional nucleic acid, other mechanisms must account for the existence of strains. The first part of this report emphasizes the role of PrP in determining strain properties, followed by discussion of other components that may be involved.

2 Incubation Time Properties of Mouse Prion Strains

2.1 The Prion Protein Gene and Scrapie Incubation Time

Although the isolation and characterization of most mouse scrapie isolates predated the discovery of PrP (BOLTON et al. 1982) and its chromosomal *Prnp* gene (OESCH et al. 1985; SPARKES et al. 1986), incubation time profiles in mice with either or both alleles of this single host gene, previously known as *Sinc,* formed the chief criteria for distinguishing scrapie strains (DICKINSON and FRASER 1977). The predominant influence of this single gene on mouse scrapie incubation time was described over 30 years ago (DICKINSON and MACKAY 1964; DICKINSON et al. 1968), but the gene's chromosomal location remained unknown until demonstration of linkage with the prion protein locus (CARLSON et al. 1986). It is now almost certain that *Prnp* itself, rather than a linked locus, controls scrapie incubation period. The *a* and *b* alleles of *Prnp* differ at codons 108 (Leu/Phe) and 189 (Thr/Val; WESTAWAY et al. 1987). The term "allotype" is used here to distinguish between the proteins encoded by the *a* and *b* alleles of *Prnp*, reserving the term "isoform" to distinguish between PrP^{C} and PrP^{Sc}. For example, PrP^{Sc}-B indicates the scrapie isoform of $Prnp^{b}$ encoded protein. The influence of each *Prnp* allele on prion incubation time depends on the scrapie strain that is inoculated. Prolongation of incubation times for the Rocky Mountain Lab (RML) isolate by the *b* allele appears to be a dominant trait with *a/a* homozygotes becoming ill at the latest 150 days after inoculation while *a/b* and *b/b* mice show no signs of disease until at least 200 days. In contrast, the *a* allele appears to be overdominant in prolonging 22A incubation period; $Prnp^{b}$ homozygous mice have incubation times of approximately 200 days and *a* homozygous animals roughly 400 days, while F1 hybrids have longer incubation times (~500 days) than either parent. Isolate-specific patterns of allelic interaction with the prion incubation time gene were taken as evidence that the scrapie agent had a genome independent of the host (BRUCE and DICKINSON 1987; DICKINSON and MEIKLE 1971). Sheep also show similar changes in allelic interaction between the prion protein gene and scrapie isolate (GOLDMAN et al. 1990; WESTAWAY et al. 1994b).

2.2 Stability of Prion Isolate Properties: Host Direction vs. Host Selection

Changes in prion properties can be interpreted in classical microbiological terms as nucleic acid mutations, but a functional host-independent macromolecule has not been detected in prions and may not be necessary to account for prion diversity. Passage history can have a profound influence on prion properties. As noted above, 22A has a relatively short 200-day incubation time in mice homozygous for the *b* allele of *Prnp* while *Prnp*^{a} mice have intervals of about 400 days between inoculation and illness. Although this incubation time profile is stable if 22A is maintained in *Prnp*^{b} homozygotes, only PrP^{Sc}-B 22A prions produce long incubation times in *Prnp*^{a} mice. Incubation time rapidly shortens through subsequent passages through mice expressing PrP-A (Bruce and Dickinson 1987). Under the microbiological paradigm, the shortening of incubation period reflects host selection for more rapidly replicating prion mutants present in the 22A isolate. However, an alternative explanation invoking preference for interaction between homologous PrP allotypes is also viable.

The first evidence suggesting the importance of PrP allotype matching was the shorter incubation time in *Prnp*^{b} homozygous mice produced by RML PrP^{Sc}-B prions (passed through mice homozygous for the *b* allele of *Prnp*) than by RML prions comprised of PrP^{Sc}-A (Carlson et al. 1989). The interval between inoculation and illness for homologous prions was shortened by over 100 days (from more than 300 to approximately 200 days), but it should be stressed that incubation periods for RML prions in *Prnp*^{b} mice never became as short as in *a/a* homozygotes regardless of passage history. A more dramatic demonstration of the effect of PrP homology between donor and host is the species barrier to scrapie transmission (Pattison and Millson 1961). Few mice inoculated with Syrian hamster (sHa) prions become ill, and those that do only after 600 days or more. Expression of sHaPrP transgenes overcomes the species barrier with incubation times as short as 50 days in high copy number mice (Scott et al. 1989; Prusiner et al. 1990). Transgenic (Tg) mice expressing sHaPrP inoculated with sHa prions produce only sHa prions while mouse prion inoculation elicits production of mouse but not sHa scrapie agent, providing evidence that interaction between PrP^{C} and PrP^{Sc} is an essential feature of prion replication.

2.3 Prion Strain Properties Reflect Interactions Between PrP^{C} and PrP^{Sc}

Freedom from the conventional constraint of three possible combinations for two alleles, provided by the availability of Tg and gene ablated mice, spurred new approaches to understanding prion strains and reevaluation of old concepts. For nearly 30 years the term "dominant" was used to describe the effects of the *b* allele of *Prnp* in prolonging scrapie incubation time (Dickinson et al. 1968; Carlson et al. 1986). For this reason and because of strong genetic evidence that *Prnp* rather than a distinct locus, controlled prion incubation period (WESTAWAY et al. 1987; CARLSON et al. 1988, 1993), we expected that expression of *Prnp*^{b} transgenes would prolong the RML incubation time of *Prnp*^{a} homozygous mice. In fact, Tg(*Prnp*^{b}) mice had shorter incubation times than their non-Tg littermates (Table 1, lines 6–9), a finding we considered to be paradoxical (WESTAWAY et al. 1991). One interpretation of this result was that a distinct incubation time locus mitigating the effect of transgene overexpression was not included within the 40 kb *Prnp*-containing cosmid insert used to produce the mice. As one test for the hypothesis of an incubation time locus separable from *Prnp*, we determined the influence of an "authentic" *Prnp*^{b} allele (or a tightly linked gene) on RML isolate incubation time in Tg15 mice. Tg(*Prnp*^{b}) 15 mice have three copies of the *Prnp*^{b} containing cosmid insert derived from I/LnJ mice. Tg 15 mice hemizygous for the transgene array were crossed with B6.I-1 (B6.I-*Prnp*^{b}/Co) congenic mice and inoculated with RML isolate. B6.I mice have long RML incubation times and are homozygous for I/LnJ-derived alleles of *Prnp* and loci distal to the G protein receptor gene *Gpcr8* and proximal to the homeobox gene *Pax-1* (CARLSON et al. 1993); other loci are derived from the short incubation time B6 strain. All offspring of this cross are *Prnp* heterozygous, with half expected to be transgene-negative and half hemizygous for the three-copy transgene array (Table 1, line 4 vs. line 5). *Prnp*^{a} homozygous mice hemizygous for the transgene array had incubation times of 115 ± 3 days compared to 144 ± 5 days for transgene-negative *Prnp*^{a} mice. In comparison, transgene-positive *Prnp*^{a}/*Prnp*^{b} heterozygous mice had incubation times of 166 ± 2 days, not as long as "authentic" *Prnp*^{a}/*Prnp*^{b} mice but a significant prolongation due to the host genotype.

These results should have been expected based the reevaluation of genetic control of scrapie incubation time prompted by our recent results using mice carrying various numbers of *a* and *b* *Prnp* alleles as summarized in Table 1. Note that mice carrying one normal *Prnp* allele and one nonfunctional allele disrupted through homologous recombination have very long incubation times. The appar-

Table 1. Influence of *Prnp* allele copy number on incubation times for RML prion isolate

Line no.	*Prnp* Copy number[a] *a* allele	*b* allele	Incubation time (days)	Number of mice
1	0	0	No Illness	
2	1	0	426 ±18	9
3	0	2	360 ±16	7
4	1	1	268 ± 4	7
5	1	4	166 ± 2	11
6	2	0	144 ± 5	8
7	2	3	115 ± 3	18
8	2	6	111 ± 5	5
9	2	>30	75 ± 2	15
10	>30	0	50 ± 2	16

[a]Copy number can exceed two when *Prnp* transgenic mice are used. Tg(*Prnp*^{b})15 mice harbor three copies of the transgene and therefore mice with two copies of the *a* allele and three copies of the *b* alleles or one copy of the *a* allele and four copies can be produced through simple crosses. Detailed reports of these experiments can be found in CARLSON et al. (1993, 1994b) and WESTAWAY et al. (1991).

ent dominance of the *b* allele in prolonging RML incubation time in *Prnp* heterozygous mice actually reflects the reduced amount of PrP^C produced from a single copy of *Prnp*^{a}. Based on these and similar results, we suggest that RML prions specify more rapid conversion of PrP^C-A into PrP^{Sc}-A than PrP^C-B to PrP^{Sc}-B. Presumably, transgene expression in *Prnp*^{a} homozygous mice has no effect on the supply of PrP^C-A, and additional PrP^C, even the less efficiently converted PrP^C-B allotype, can only shorten incubation time. Note that incubation times in transgene negative heterozygous mice (Table 1, line 4) are longer than transgene positive (Table 1, line 5) mice due to the expression of four rather than one *Prnp*^{b} gene. A plot of incubation time against "corrected" *Prnp* copy number illustrates that both PrP-A and PrP-B influence incubation time but that the PrP-A allotype has a stronger effect than the B allotype (Fig. 1). In other words, the supply of PrP^C-A rather than an active dominant effect of PrP^C-B, is the primary determinant of RML incubation time (Carlson et al. 1994a,b).

Incubation times for the 22A scrapie isolate are shorter in *Prnp*^{b} mice than in *Prnp*^{a} animals. In further contrast with Chandler and other commonly used isolates, *Prnp* heterozygous mice have longer incubation times than either parent, as illustrated by incubation times in B6 (405 ± 2 days), B6.I (194 ± 10 days), and their F1 hybrid (508 ± 14 days; Fig. 2). Incubation times in Tg15 mice revealed that increasing the amount of PrP-B shortened incubation time. Transgene homozygous Tg15(+/+) and hemizygous Tg15(+/0) were inoculated along with the congenic strains. Hemizygous Tg15(+/0) mice had much shorter incubation times than "authentic" *Prnp*^{a}/*Prnp*^{b} heterozygotes (395 ± 12 vs.508 ± 14 days), and incubation times were further reduced by transgene homozygosity (286 ± 15 days). It is important to stress that incubation times of Tg15(+/+) mice with six copies of *Prnp*^{b} and two copies of *Prnp*^{a} were longer than those of B6.I mice with

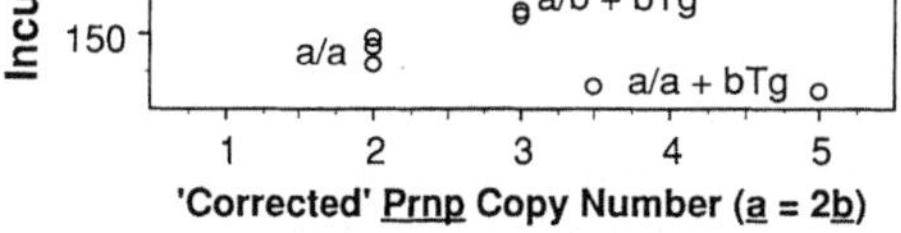

Fig. 1. Influence of *Prnp* copy number on RML prion incubation time. The effect of the *a* allele is stronger than the *b* allele. *Each circle*, an experiment with multiple mice; some of these data also are shown in Table 1. The genotype of mice in each group is given adjacent to the circle; 0, an ablated allele, *'bTg'*, the three-copy *Prnp*^{b} transgene array from Tg15 mice. Incubation time (days) cannot be related soley to *a* or *b* copy number nor to total *Prnp* copies. "Corrected" copy number where the effect of the *a* allele is assumed to be equal to twice that of the *b* allele is shown (*x axis*) to illustrate that both alleles influence incubation time, but that the *a* allele has a stronger effect. This mathematical assumption is for illustrative purposes only, and we do not imply than "*a* =2*b*" is the actual relationship

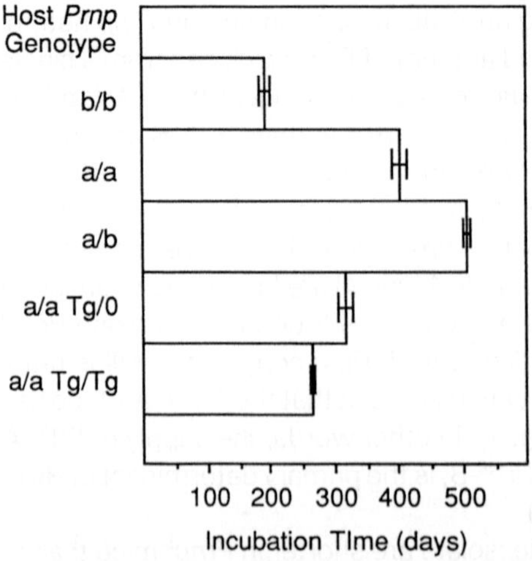

Fig. 2. Increasing copy number of *Prnp*^{b} mitigates the effect of the *a* allele in prolonging incubation times for the 22A prion strain. C57BL/6J (*Prnp*^{a}), B6.I–1 (*Prnp*^{b}), Tg(*Prnp*^{b}), and crosses among them were used these experiments. *Tg*, the three-copy transgene array from Tg (*Prnp*^{b}) 15 mice; 0, a wild-type insertion site. See also CARLSON et al. (1994b) for experimental details

two copies of *Prnp*^{b}. This result indicates that prolongation of 22A incubation time is a dominant, positive result of *Prnp*^{a} expression.

Even though results using Tg15 mice to dissect 22A isolate properties are dramatically different from those obtained with the Chandler isolate, we propose that they also can be explained by allotype-dependent PrP^C-PrP^{Sc} interactions. PrP^{Sc} in 22A prions converts PrP^C-B to PrP^{Sc} more efficiently than it does PrP^C-A, resulting in shorter incubation times for B6.I than B6 mice. An interesting feature of the incubation time profile for 22A is the overdominance of the *a* allele of *Prnp* in prolonging incubation period. On the basis of this overdominance, DICKINSON and OUTRAM (1979) proposed a replication site hypothesis, postulating that dimers of the *Sinc* gene product feature in the replication of the scrapie agent. The results in Fig. 2 are compatible with the interpretation that the target for PrP^{Sc} may be a PrP^C dimer or multimer. The assumptions under this model are that PrP^C-B dimers are more readily converted to PrP^{Sc} than are PrP^C-A dimers, and that PrP^C-A: PrP^C-B heterodimers are very resistant to conversion to PrP^{Sc}. Increasing the ratio of PrP-B to PrP-A would lead to shorter incubation times by favoring the formation of PrP^C-B homodimers. A homologous mechanism may account for the relative paucity of individuals heterozygous for the Met/Val polymorphism at codon 129 of the human PrP gene in spontaneous Creutzfeldt-Jakob disease (CJD; PALMER et al. 1991).

Alternatively, PrP^C-PrP^{Sc} interaction can be broken down to two distinct aspects, binding affinity and susceptibility to conversion. If PrP-A has a higher affinity for 22A PrP^{Sc} than does PrP^C-B, but is inefficiently converted to PrP^{Sc}, the exceptionally long incubation time of *Prnp* heterozygotes might reflect reduction in the supply of 22A prions available for interaction with the PrP^C-B product of the single *Prnp*^{b} allele. Results using the 87V prion isolate provide evidence for high-affinity PrP^C-PrP^{Sc} interactions in the absence of conversion to PrP^{Sc} and favor the possibility that binding affinities and rate constants may be sufficient to explain incubation time profiles of prion strains.

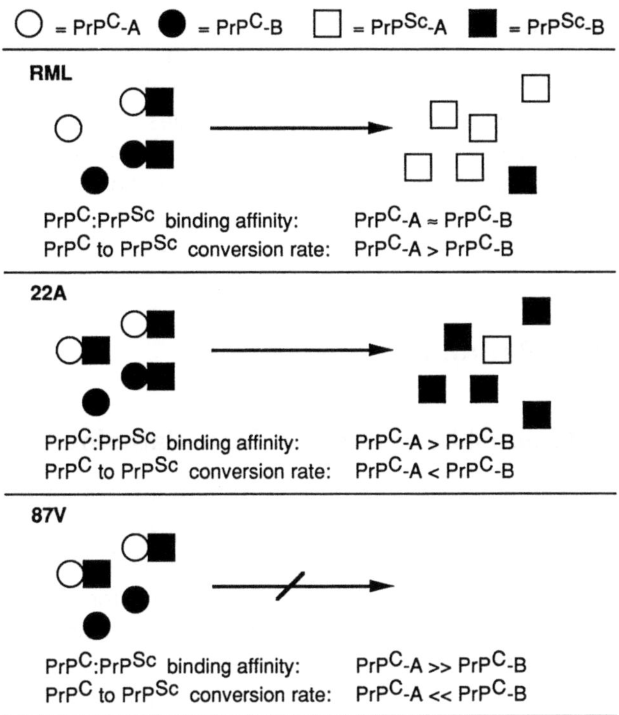

Fig. 3. Model for interactions between PrP^{C} and PrP^{Sc} to explain incubation time properties of prion strains. The three panels illustrate *Prnp* heterozygous mice inoculated with RML, 22A, or 87V prion strains. See text for detailed discussion

Of *Prnp*^{a}, *Prnp*^{b}, *Prnp*^{a}/*Prnp*^{b}, and Tg15 mice inoculated with 87V prions, only *Prnp*^{b} mice became uniformly ill. Only a single NZW/LacJ of 31 inoculated was diagnosed with scrapie confirmed by pathological analysis. Unlike the results with other scrapie isolates, expression of the three-copy *Prnp*^{b} transgene array did not alter scrapie susceptibility. The postulate of a high binding affinity of PrP^{C}-A for 87V PrP^{Sc}-B without conversion to PrP^{Sc}-A is sufficient to explain these results. Figure 3 summarizes how prion-specified, allotype-dependent affinities of PrP^{Sc} for PrP^{C} and rate constants for PrP^{C} conversion provide a working model for incubation times of diverse scrapie strains.

It is clear the amino acid sequence of host PrP^{C} places major constraints on disease manifestation and incubation time, as does that of PrP^{Sc} in the inoculum. For example, the pathological changes produced by the RML isolate (composed of PrP^{Sc}-A) in *Prnp*^{a} and *Prnp*^{b} congenic mice are not identical, with RML-induced pathology in B6.I (*Prnp*^{b}) mice more similar to that induced by 22A PrP^{Sc}-B prions than to disease caused by RML in B6 mice (*Prnp*^{a}; Carlson et al. 1994a,b). It is important to stress that although PrP allotype places constraints on prion properties, additional levels of complexity, either higher order structure or cofactors, are necessary for prion diversity. For example, both PrP^{Sc}-A and PrP^{Sc}-B are permissive for the RML prion-specified property of shorter incubation times in *Prnp*^{a} mice than *Prnp*^{b} animals. In other words, even though interactions between homologous allotypes favor shorter incubation times (CARLSON et al. 1989), the incubation time profile and induced pathological changes are stable regardless of passage history. In contrast, *Prnp*^{a} mice appear unable to replicate the properties specified by 22A or 87V prions, suggesting that PrP-B is able to adopt conformations or bind cofactors not available to PrP-A.

3 Prion Strain Specific Pathology

3.1 Prion Strains and Regional Differences in Brain Pathology

"Lesion profiles" for the degree of spongiform degeneration in different brain regions and the presence or absence of amyloid plaques provided additional criteria for distinguishing among scrapie isolates (DICKINSON and FRASER 1977; BRUCE et al. 1976, 1991). Genes of the host, as well as prion-specified information, determine the character of PrP^{Sc} accumulation and attendant neuropathology. A single isolate, such as RML, can produce distinct pathologies in *Prnp*^{a} and *Prnp*^{b} mice. Even more dramatic evidence for the influence of PrP primary structure on pathology (and clinical presentation) comes from human familial prion diseases. Each PRNP mutation is associated with a relatively set constellation of pathology and clinical signs, although non-PRNP genes can produce intrafamily variation. The most clearcut case of the impact of PrP amino acid sequence on disease phenotype is the effect of the nonpathogenic Met/Val codon 129 polymorphism in combination with the pathogenic 178Asn mutation. When Asn-178 is inherited in *cis* with Val-129 fatal familial insomnia (FFI) results while the Met-129 – Asn-178 allele produces CJD (GOLDFARB et al. 1992).

Mouse isolates 22A, 87V and RML produce distinctive patterns of PrP^{Sc} accumulation and pathology in the same inbred mouse strain (Fig. 4). 87V inoculation leads to the development of amyloid plaques. PrP^{Sc} deposition in B6.I-1 mice inoculated with 87V prions was most intense in the thalamus (particularly in the ventral posterior lateral nucleus), habenula, and locus ceruleus and raphe nuclei of the brainstem. Little or no 87V PrP^{Sc} accumulated in the neocortex, hippocampus, or hypothalamus. In contrast, 22A isolate produces PrP^{Sc} staining in the full thickness of the cortex, granule cell layer of the cerebellum, hippocampus, and hypothalamus. No amyloid plaques were detected in B6.I-1 mice inoculated with 22A. Interestingly, RML in B6.I-1 mice produces pathology more similar to 22A than to RML in *Prnp*^{a} C57BL/6J mice. Although the general features of PrP^{Sc} deposition are shared between 22A and RML, clear differences are seen. For example, there is more intense PrP^{Sc} immunostaining in the molecular layer of the dentate gyrus with RML compared to the presence of more PrP^{Sc} in the hippocampus of 22A inoculated mice. In addition, the full thickness of the neocortex was

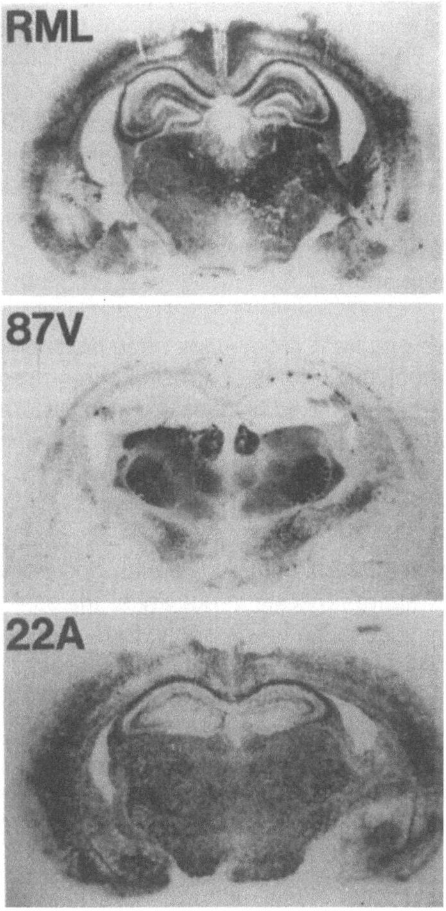

Fig. 4. Patterns of PrP^{Sc} accumulation differ for scrapie strains RML, 87V, and 22A in the same inbred mouse strain B6.I-1. Immunostaining of proteinase K treated histoblots (TARABOULOS et al. 1992) of brain sliced at the level of the thalamus and hippocampus reveals distinctive patterns of PrP^{Sc} accumulation for each prion strain. These histoblots were included as a portion of a more comprehensive figure in CARLSON et al. (1994b)

involved by 22A prions, but only the inner layers contained RML PrP^{Sc}. Although there is suggestive evidence for electrophoretic mobility differences in PrP between prion isolates (BESSEN and MARSH 1992; KASCSAK et al. 1986), presumably including differences in glycosylation, the relevance of these differences to prion properties is unknown.

3.2 Host Genes in Addition to *Prnp* Influence Prion-Induced Pathology

A novel approach towards understanding the molecular bases for isolate-specific differences in patterns of pathology came from work using *Prnp* congenic mice (CARLSON et al. 1994b). I/LnJ mice inoculated with 87V prions show little deposition of PrP^{Sc} and vacuolation in the thalamus or in the locus ceruleus and raphe nuclei of the brainstem. As noted above, B6.I-1 mice have intense pathology in these regions. B6.I-1 mice contain a small segment of I/LnJ-derived chromosome 2 on a C57BL/6J background. Another congenic strain, designated B6.I-2, has 87V-induced pathological changes identical to those in the I/LnJ donor strain, suggesting that an allele of a locus present in B6.I-2 but absent in B6.I-1 suppresses the deposition of PrP^{Sc} and thus vacuolation in the thalamus, habenula, and raphe nuclei. Because these strains are congenic, it is likely that this novel locus is linked to *Prnp,* but segregation analysis is necessary to confirm this. RFLP typing using end probes from Cos6.I/LnJ-4 (Westaway et al. 1994a) indicated that the 40 kb *Prnp*-containing segment is I/LnJ derived, excluding intragenic recombination.

Although host genetic influences on prion pathology have been described previously, for the most part they were polygenic traits controlling differences between inbred strains of mice. For example, amyloid plaques are a consistent pathological feature of 87V prions in VM/Dk mice (as well as in I/LnJ, B6.I-1, and B6.I-2) but are absent from IM/Dk (Bruce and Dickinson 1985). A candidate for a single gene influencing the pathological changes induced by prion inoculation is the mink homologue of the mouse beige (*bg*) mutation and the mutation causing Chediak-Higashi disease in humans (Marsh et al. 1976). Aleutian mink inoculated with transmissible mink encephalopathy agent replicate infectivity at the same rate as wild-type animals and have similar incubation times; spongiform change, however, is minimal in Aleutian mink. This apparent dissociation of incubation time and prion replication from pathology also has implications for understanding prion diseases and may relate to the intracellular membrane defect affecting lysosomal and granule function in animals with Chediak-Higashi disease. The relationship between microscopic pathology and neurological dysfunction is poorly understood, but a wide spectrum of pathologies and symptomology are seen in human prion diseases.

4 Overview

Use of transgenic mice to dissect incubation time profiles of prion strains produced results that could be explained solely by affinity of PrP^{C}-PrP^{Sc} binding and the rate constant of subsequent conversion of PrP^{C} to PrP^{Sc}. We do not suggest that our results prove that PrP^{Sc} is the sole functional component of prions. However, we have developed a single model that can account for the incubation time properties of a variety of prion strains without invoking an additional macromolecule that encodes information. These studies do not directly address differences in pathology produced by distinct scrapie isolates, but these also may not demand a second component. FFI and CJD are clinically and pathologically distinct prion disorders, yet both have been linked to an identical mutation at codon 178 of PRNP (Goldfarb et al. 1992). The major determinant of which disease results from the Asn-178 mutation is the Met/Val polymorphism at codon 129. If the Asn-178 mutation is present on the Met-129 allele, FFI occurs, while the same mutation on the Val-129 allele produces CJD. Recent analysis of PrP^{Sc} from FFI and Asn-178 CJD revealed differences in the relative abundance of glycosylated forms and in the size of protease resistant fragments (MONARI et al. 1994). The size difference may be explained by a change in protease cleavage sites resulting from a different conformation of PrP^{Sc} in the two diseases. Similarly, chimeric PrP molecules constructed from mouse and hamster genes produce prions with novel properties (SCOTT et al. 1993).

However, scrapie isolates 22A and 87V, for example, both consist of PrP^{Sc}-B, and clearly exhibit different properties. These differences might reflect alternative conformations of PrP^{Sc} or distinct covalent posttranslational modifications. Different cell types in the brain may produce different isoforms of PrP^{C}, which may account for specific targeting of scrapie strains (DEARMOND et al. 1993). Definitive determination of the biochemical basis for scrapie strain behavior will likely result from an understanding of the nature of interaction between the cellular and scrapie isoforms of PrP and the physical differences between the two isoforms.

Acknowledgments. Thanks are due to Drs. Stanley B. Prusiner, Steven J. DeArmond, David Westaway, Shu-lian Yang, Michael Scott, Marilyn Torchia, Darlene Groth, Glen Telling and other members of the University of California, San Francisco scrapie group whose scientific contributions and insight formed the basis for this review.

References

Bessen RA, Marsh RF (1992) Biochemical and physical properties of the prion protein from two strains of the transmissible mink encephalopathy agent. J Virol 66: 2096–2101

Bolton DC, McKinley MP, Prusiner SB (1982) Identification of a protein that purifies with the scrapie prion. Science 218: 1309–1311

Bruce ME, Dickinson AG (1985) Genetic control of amyloid plaque production and incubation period in scrapie-infected mice. J Neuropathol Exp Neurol 44: 285–294

Bruce ME, Dickinson AG (1987) Biological evidence that the scrapie agent has an independent genome. J Gen Virol 68: 79–89

Bruce ME, Dickinson AG, Fraser H (1976) Cerebral amyloidosis in scrapie in the mouse: effect of agent strain and mouse genotype. Neuropathol Appl Neurobiol 2: 471–478

Bruce ME, McConnell I, Fraser H, Dickinson AG (1991) Disease characteristics of different strains of scrapie in *Sinc* congenic mouse lines: implications for the nature of the agent and host control of pathogenesis. J Gen Virol 72: 595–603

Carlson GA, Kingsbury DT, Goodman PA, Coleman S, Marshall ST, DeArmond S, Westaway D, Prusiner SB (1986) Linkage of prion protein and scrapie incubation time genes. Cell 46: 503–511

Carlson GA, Goodman PA, Lovett M, Taylor BA, Marshall ST, Peterson-Torchia M, Westaway D, Prusiner SB (1988) Genetics and polymorphism of the mouse prion gene complex: control of scrapie incubation time. Mol Cell Biol 8: 5528–5540

Carlson GA, Westaway D, DeArmond SJ, Peterson-Torchia M, Prusiner SB (1989) Primary structure of prion protein may modify scrapie isolate properties. Proc Natl Acad Sci USA 86: 7475–7479

Carlson GA, Ebeling C, Torchia M, Westaway D, Prusiner SB (1993) Delimiting the location of the scrapie prion incubation time gene on chromosome 2 of the mouse. Genetics 133: 979–988

Carlson GA, DeArmond SJ, Torchia M, Westaway D, Prusiner SB (1994a) Genetics of prion diseases and prion diversity in mice. Philos Trans R Soc Lond [B] 343: 363–369

Carlson GA, Ebeling C, Yang SL, Telling G, Torchia M, Groth D, Westaway D, DeArmond SJ, Prusiner SB (1994b) Prion isolate specified allotypic interactions between the cellular and scrapie prion proteins in congenic and transgenic mice. Proc Natl Acad Sci USA 91: 5690–5694

DeArmond SJ, Yang S-L, Lee A, Bowler R, Taraboulos A, Groth D, Prusiner SB (1993) Three scrapie prion isolates exhibit different accumulation patterns of the prion protein scrapie isoform. Proc Natl Acad Sci USA 90: 6449–6453

Dickinson AG, Fraser H (1977) Scrapie: pathogenesis in inbred mice: an assessment of host control and response involving many strains of agent. In: Katz M, Meuler V (eds) Slow virus infections of CNS. Springer, Berlin Heidelberg New York, pp 3–14

Dickinson AG, MacKay JMK (1964) Genetical control of the incubation period in mice of the neurological disease, scrapie. Heredity 19: 279–288

Dickinson AG, Meikle VMH (1971) Host-genotype and agent effects in scrapie incubation: change in allelic interaction with different strains of agent. Mol Gen Genet 112: 73–79

Dickinson AG, Outram GW (1979) The scrapie replication-site hypothesis and its implications for pathogenesis. In: Prusiner SB, Hadlow WJ (eds) Slow transmissible diseases of the nervous system, vol 2. Academic, New York, pp 13–31

Dickinson AG, Meikle VMH, Fraser HG (1968) Identification of a gene which controls the incubation period of some strains of scrapie agent in mice. J Comp Pathol 78: 293–299

Goldfarb LG, Petersen RB, Tabaton M, Brown P, LeBlanc AC, Montagna P, Cortelli P, Julien J, Vital C, Pendelbury WW, Haltia M, Wills PR, Hauw JJ, McKeever PE, Monare L, Schrank B, Swergold GD, Autilio-Gambetti L, Gajdusek DC, Lugaresi E, Gambetti P (1992) Fatal familial insomnia and familial Creutzfeldt-Jakob disease: disease phenotype determined by DNA polymorphism. Science 258: 806–808

Goldman W, Hunter N, Foster JD, Salbaum JM, Beyreuther K, Hope J (1990) Two alleles of a neural protein gene linked to scrapie in sheep. Proc Natl Acad Sci USA 87: 2476–2480

Kascsak RJ, Rubenstein R, Merz PA, Carp RI, Robakis NK, Wisniewski HM, Diringer H (1986) Immunological comparison of scrapie-associated fibrils isolated from animals infected with four different scrapie strains J Virol 59: 676–683

Kellings K, Meyer N, Mirenda C, Prusiner SB, Riesner D (1992) Further analysis of nucleic acids in purified scrapie prion preparations by improved return refocussing gel electrophoresis (RRGE). J Gen Virol 73: 1025–1029

Marsh RF, Sipe JC, Morse SS, Hanson RP (1976) Transmissible mink encephalopathy reduced spongiform degeneration in aged mink of the Chediak-Higashi genotype. Lab Invest 34: 381–386

Monari L, Chen SG, Brown P, Parchi P, Petersen RB, Mikol J, Gray F, Cortelli P, Montagna P, Ghetti B, Goldfarb LG, Gajdusek DC, Lugaresi E, Gambetti P, Autilio-Gambetti L (1994) Fatal familial insomnia and familial Creutzfeldt-Jakob disease: different prion proteins determined by a DNA polymorphism. Proc Natl Acad Sci USA 91: 2839–2842

Oesch B, Westaway D, Wächli M, McKinley MP, Kent SBH, Aebersold R, Barry RA, Tempst P, Teplow DB, Hood L, Prusiner SB, Weissmann C (1985) A cellular gene encodes scrapie PrP 27–30 protein. Cell 40: 735–746

Palmer MS, Dryden AJ, Hughes JT, Collinge J (1991) Homozygous prion protein genotype predisposes to sporadic Creutzfeldt-Jakob disease. Nature 352: 340–342

Pan K-M, Baldwin M, Nguyen J, Gasset M, Serban A, Groth D, Melhorn I, Huang Z, Fletterick RJ, Cohen FE, Prusiner SB (1993) Conversion of α-helices into β-sheets features in the formation of the scrapie prion proteins. Proc Natl Acad Sci USA 90: 10962–10966

Pattison IH, Millson GC (1961) Scrapie produced experimentally in goats with special reference to the clinical syndrome. J Comp Pathol 71: 101–108

Prusiner SB (1991) Molecular biology of prion diseases. Science 252: 1515–1522

Prusiner SB, Scott M, Foster D, Pan K-M, Groth D, Mirenda C, Torchia M, Yang S-L, Serban D, Carlson GA, Hoppe PC, Westaway D, DeArmond SJ (1990) Transgenetic studies implicate interactions between homologous PrP isoforms in scrapie prion replication. Cell 63: 673–686

Safar J, Roller PP, Gajdusek DC, Gibbs CJ Jr. (1993) Conformation transitions, dissociation, and unfolding of scrapie amyloid (prion) protein. J Biol Chem 269: 20276–20284

Scott M, Foster D, Mirenda C, Serban D, Coufal F, Wächli M, Iorchia M, Groth D, Carlson G, DeArmond SJ, Westaway D, Prusiner SB (1989) Transgenic mice expressing hamster prion protein produce species- specific scrapie infectivity and amyloid plaques. Cell 59: 847–857

Scott M, Groth D, Foster D, Torchia M, Yang S-L, DeArmond SJ, Prusiner SB (1993) Propagation of prions with artificial properties in transgenic mice expressing chimeric PrP genes. Cell 73: 979–988

Sparkes RS, Simon M, Cohn VH, Fournier REK, Lem J, Klisak I, Heinzmann C, Blatt C, Lucero M, Mohandas T, DeArmond SJ, Westaway D, Prusiner SB, Weiner LP (1986) Assignment of the human and mouse prion protein genes to homologous chromosomes. Proc Natl Acad Sci USA 83: 7358–7362

Taraboulos A, Jendroska K, Serban D, Yang S-L, DeArmond SJ, Prusiner SB (1992) Regional mapping of prion proteins in brains. Proc Natl Acad Sci USA 89: 7620–7624

Westaway D, Goodman PA, Mirenda CA, McKinley MP, Carlson GA, Prusiner SB (1987) Distinct prion proteins in short and long scrapie incubation period mice. Cell 51: 651–662

Westaway D, Mirenda CA, Foster D, Zebarjadian Y, Scott M, Torchia M, Yang SL, Serban H, DeArmond SJ, Ebeling C, Prusiner SB, Carlson GA (1991) Paradoxical shortening of scrapie incubation times by expression of prion protein transgenes derived from long incubation time mice. Neuron 7: 59–68

Westaway D, Cooper C, Turner S, Da Costa M, Carlson GA, Prusiner SB (1994a) Structure and polymorphism of the mouse prion protein gene. Proc Natl Acad Sci USA 91: 6418–6422

Westaway D, Zuliani V, Cooper CM, Da Costa M, Neuman S, Jenny AL, Detwiler L, Prusiner SB (1994b) Homozygosity for prion protein alleles encoding glutamine-171 renders sheep susceptible to natural scrapie. Genes 8: 959–969

Wickner RB (1994) [URE3] as an altered *URE2* protein: evidence for a prion analog in *Saccharomyces cerevisiae.* Science 264: 566–569

Zlotnik I (1965) Observations on the experimental transmission of scrapie of various origins to laboratory animals. In: Slow, latent, and temperate virus infections. NINDB Monogr 2: 237–248

Structures of Prion Proteins and Conformational Models for Prion Diseases

Z. HUANG[1,2,5], S.B. PRUSINER[2,3], and F.E. COHEN[1,3,4]

1 Introduction

Prions are a novel class of "infectious" pathogens distinct from viroids and viruses with respect to both their structure and the neurodegenerative diseases that they cause (PRUSINER 1991). Prion diseases are manifest as sporadic, inherited, and infectious disorders including scrapie, mink encephalopathy, chronic wasting disease, bovine spongiform encephalopathy, feline spongiform encephalopathy, and exotic ungulate encephalopathy of animals (MARSH et al. 1991; WESTAWAY et al. 1994b; WILESMITH and WELLS 1991) as well as kuru, Creutzfeldt-Jakob disease (CJD), Gerstmann-Sträussler-Scheinker syndrome, and fatal familial insomnia of humans (GAJDUSEK 1977; HSIAO et al. 1989; MEDORI et al. 1992). The prion protein (PrP) is the major, if not the only, component of prions (PRUSINER 1991). PrP exists in two isoforms: the normal cellular form (PrP^C) and the abnormal disease-related form (PrP^{Sc}) (MEYER et al. 1986; OESCH et al. 1985).

[1] Department of Pharmaceutical Chemistry, University of California, San Francisco, CA 94143, USA
[2] Department of Neurology, University of California, San Francisco, CA 94143, USA
[3] Department of Biochemistry and Biophysics, University of California, San Francisco, CA 94143, USA
[4] Department of Medicine, University of California, San Francisco, CA 94143, USA
[5] Present Address: Jefferson Cancer Institute, Thomas Jefferson University, Philadelphia, PA 19107, USA

The entire open reading frame of PrP genes is contained in a single exon, eliminating the possibility that PrP^C and PrP^{Sc} arise from alternative RNA splicing (BASLER et al. 1986). Attempts to identify a posttranslational chemical modification that features in the conversion of PrP^C to PrP^{Sc} have been unsuccessful (STAHL et al. 1993). Structural studies of PrP^C and PrP^{Sc} using Fourier-transform infrared (FTIR) and circular dichroism (CD) spectroscopy indicated that PrP^C and PrP^{Sc} differ from each other in their conformations (CAUGHEY et al. 1991; GASSET et al. 1993; PAN et al. 1993; PRUSINER et al. 1983; SAFAR et al. 1993a,b). These findings suggest that prion diseases are disorders of protein conformation and result from a change in the structure of PrP^C when it is converted into PrP^{Sc}.

Elucidating the three-dimensional structures of both PrP^C and PrP^{Sc} as well as the conformational changes that occur during the production of PrP^{Sc} are central to understanding the molecular mechanisms of prion diseases. Because of the low level of expression of PrP^C and insolubility of PrP^{Sc}, efforts to obtain crystals for X-ray structure determinations have been unsuccessful. Exploiting recent advances in protein structure prediction algorithms, we carried out computational studies to predict the three-dimensional structure of PrP^C and PrP^{Sc} based on a family of homologous amino acid sequences.

Our predictive studies were facilitated by spectroscopic findings showing that PrP^C has a secondary structure which contains approx.43% α-helix and is virtually devoid of β-sheet, wheras PrP^{Sc} has a β-sheet content of 43% and an α-helix content of 30% (PAN et al. 1993). Although the problem of predicting protein structure from the amino acid sequence alone remains unsolved, recent advances in secondary and tertiary structure prediction of proteins have shown that reasonable structures can often be proposed using a heuristic approach in conjunction with experimental data (RING and COHEN 1993). For both all α-helical proteins and α/β proteins this approach has been applied to generate low resolution structural models for a number of proteins (BOISSEL et al. 1993; COHEN et al. 1986b; CURTIS et al. 1991; HURLE et al. 1987). Some of the structural features proposed by these methods have been verified by subsequent X-ray or nuclear magnetic resonance experiments (SMITH et al. 1992).

In the study reported here three-dimensional structures of PrP^C and PrP^{Sc} were generated by applying the heuristic approach coupled with experimental data. In the absence of a three-dimensional structure of PrP^C and PrP^{Sc} from either X-ray crystallography or NMR spectroscopy these models provide a vehicle to rationalize much of the available biological and genetic data. Furthermore, they reveal a plausible structural basis for prion diseases.

2 Methods

The computational procedures used for the prediction of the three-dimensional structures of PrP^C and PrP^{Sc} involved four major steps: (a) alignment of a family of homologous sequences, (b) prediction of secondary structures, (c) packing of secondary elements to generate all plausible tertiary structures, and (d) selection and refinement of final structural models. PrP amino acid sequences from one avian and 11 mammalian sources including chicken, cow, sheep, rat, mouse, hamster, mink, and humans were used. The alignment of these sequences was reported previously (GABRIEL et al. 1992) using the FENG and DOOLITTLE (1987) algorithm. Methods for secondary and tertiary structure prediction were applied independently to all 12 PrP sequences. As recently suggested by BENNER and GERLOFF (1991), the use of a broad family of homologous sequences improves the accuracy of structure prediction.

2.1 Structure Prediction of PrP^C

An initial attempt to predict the secondary structure of PrP^C was made using several different methods including the Chou-Fasman method (CHOU and FASMAN 1978), Garnier-Osguthorpe-Robson (GOR) algorithm (GARNIER et al. 1978) and PHD program (ROST and SANDER 1992). The GOR algorithm was used without setting a decision constant (GARNIER et al. 1978). Results of the PHD method were obtained from the European Molecular Biology Laboratories. As discussed below, these general-purpose prediction methods gave confusing assignments for secondary structure. Given the experimental data indicating that PrP exists in two distinct conformations, the inconsistencies among various algorithms could reflect their intrinsic limitations in predicting one structure for a protein with two or more distinct conformational isoforms.

In an attempt to overcome the inherent limitations of secondary structure prediction methods we considered information derived from spectroscopic studies of PrP^C. FTIR and CD spectroscopy indicated that PrP^C contains 43% α-helices and virtually no β-sheet (PAN et al. 1993). These data indicated that PrP^C is a member of the structural class of proteins that are all α-helical. With this knowledge, we first applied a pattern-based turn prediction algorithm (COHEN et al. 1986a) to identify sequence regions that could contain secondary structures, and then predicted α-helices or β-strands within these regions using a neural network (KNELLER et al. 1990; PRESNELL et al. 1993). While most general-purpose secondary structure prediction algorithms achieve only approx. 65% accuracy due to the lack of tertiary information (GARNIER 1990), the turn-prediction algorithm and neural network method, having been optimized to exploit the characteristics of all α-helical protein structures, are accurate 90% and 79% of the time, respectively (COHEN et al. 1986a; KNELLER et al. 1990). Four sequence regions were identified from the turn-prediction algorithm, and these regions were subsequently predicted to contain helices by the neural network method. Independent calculations of polypeptide chain flexibility using the method of KARPLUS and SCHULZ (1985) also indicated that these regions are the least flexible in the entire PrP sequence and thus most likely to contain secondary structure.

Tertiary structures were generated by investigating the plausible arrangements of the α-helices via a combinatorial packing approach (COHEN et al. 1979;

CURTIS et al. 1991). This approach was developed based on the original concept of the packing of α-helices proposed by CRICK (1953). The general geometric features of helix-helix packing are followed in a manner compatible with X-ray data on the structures of globular proteins (CHOTHIA et al. 1977; RICHMOND and RICHARDS 1978). The algorithm is divided into three steps: (a) the program AAPATCH identifies clusters of hydrophobic residues within the putative helices that could mediate helix-helix interactions; these fall into three geometric subclassess; (b) AAFOLD generates all possible helix pairings according to the location and geometric preferences of the interaction sites; and (c) AABUILD generates three-dimensional models of all possible structures from the list of helix pairings (from AAFOLD) subject to steric restrictions and geometric constraints on chain folding. Additional physical and chemical data for PrP^C such as the experimentally determined disulfide bridge (TURK et al. 1988) and glycosylation sites as well as general features of proteins such as the expected relationship of surface area to molecular volume were used to reduce further the set of plausible alternative structures. At this stage, coordinates were specified for the core helices of the protein. Loops connecting these core helices were inserted by using the BLOOP program (RING and COHEN 1994). In general the conformations of loops and side chains in proteins are more flexible than the backbone of secondary elements; consequently, their conformations are less well defined in the proposed models.

2.2 Structure Prediction of PrP^{Sc}

The secondary structure prediction of PrP^{Sc} was carried out based on the spectroscopic data from CD and FTIR experiments. CD and FTIR studies indicated that approximately half of α-helical structure in PrP^C is converted into β-sheet in PrP^{Sc} (PAN et al. 1993; SAFAR et al. 1993a). Results from various secondary structure prediction methods and spectroscopic studies of PrP peptide fragments indicated that the four structural regions corresponding to the helices in PrP^C are most likely to undergo a conformational change from α-helix to β-sheet in PrP^{Sc} (GASSET et al. 1993; HUANG et al. 1994). This suggested that two of the four helices of PrP^C are converted into β-sheet structure in PrP^{Sc}. Therefore we systematically assigned two of the four helical regions to β-sheet conformations and thus obtained six different sets of secondary structure, i.e., $S_{1a1b}S_{2a2b}H_3H_4$, $S_{1a1b}H_2S_{3a3b}H_4$, $S_{1a1b}H_2H_3S_{4a4b}$, $H_1S_{2a2b}S_{3a3b}H_4$, $H_1S_{2a2b}H_3S_{4a4b}$, and $H_1H_2S_{3a3b}S_{4a4b}$. S_{1a1b} denotes a β hairpin structure containing two antiparallel β-strands (1a and 1b respectively) which is converted from helix 1 in PrP^C. The boundaries of the β-strands and the β turn of the hairpin were predicted by using a turn prediction method for α/β proteins (COHEN et al. 1986a).

The tertiary structures of PrP^{Sc} were generated by examining the plausible hydrophobic packing between secondary structures. A combinatorial packing approach developed for α/β proteins (COHEN et al. 1982; HURLE et al. 1987) was employed to predict plausible tertiary structural models for PrP^{Sc}. All possible β-strand arrangements and relative pairing of the component β-strands to form a β-sheet were studied. The α-helices were then placed onto the hydrophobic surface of the β-sheet structure so that a constellation of nonpolar residues on the surface of each α-helix can interact with the hydrophobic surface of the β-sheet. Structures were eliminated if their β-sheets failed to create a hydrophobic surface suitable for subsequent helix-sheet packing or created steric problems for the loops that would join the β-strands. Structures were also eliminated if they failed to form the experimentally determined disulfide bond between Cys-179 and Cys-214 (TURK et al. 1988). Finally, the generated tertiary models were examined in order to select the most plausible models based on the available experimental data.

3 Results

3.1 Three-Dimensional Structural Models of PrP^C

Twelve PrP sequences were aligned (GABRIEL et al. 1992) and analyzed using a pattern-based turn-prediction method and a neural network program. Four putative helical regions of PrP^C were identified (Fig. 1A). The different assignments of α-helices or β-strands within these four helical regions made by other methods such as the Chou-Fasman and GOR algorithms are also shown (Fig. 1A). Ten of 11 point mutations that segregate with the inherited prion diseases (PRUSINER and HSIAO 1994) lie either within or immediately adjacent to these four regions of secondary structure (Fig. 1B). These observations are consistent with the findings presented here which argue that the four predicted α-helices of PrP^C are structurally and functionally important regions.

The combinatorial helix-helix packing approach was applied to search for plausible tertiary structures of PrP^C. Within the four predicted helical regions 15 residues were identified as potential sites that could mediate helix-helix interactions. Table 1 shows the location of these amino acid residues and their relative conservation over 12 PrP sequences. Except for residue 184 in helix 3, all potential helix-helix interaction residues were highly conserved. Even though residue 184 is more variable, further analysis revealed that either Ile or Val at this position can maintain the helix-helix interaction at this site (RICHMOND and RICHARDS 1978). An exhaustive search of plausible packing arrangements of four helices mediated by their possible helix-helix interaction sites gave rise to a total of approx. 300 000 structures. Among this large number of structures only a small number of structures were sterically reasonable, with a sufficient number of residues in the loops between these helices to create a stereochemically sensible structure, and still form the experimentally determined disulfide bond between Cys-179 in helix 3 and Cys-214 in helix 4 (TURK et al. 1988). These structures were clustered into approx. 200 structural families. The remaining structures were then rank ordered by solvent accessible surface area. Four structures were chosen as the plausible models of PrP^C (Fig. 2). These structures were significantly more compact than the rest as judged by the ratio of accessible surface area to molecular volume.

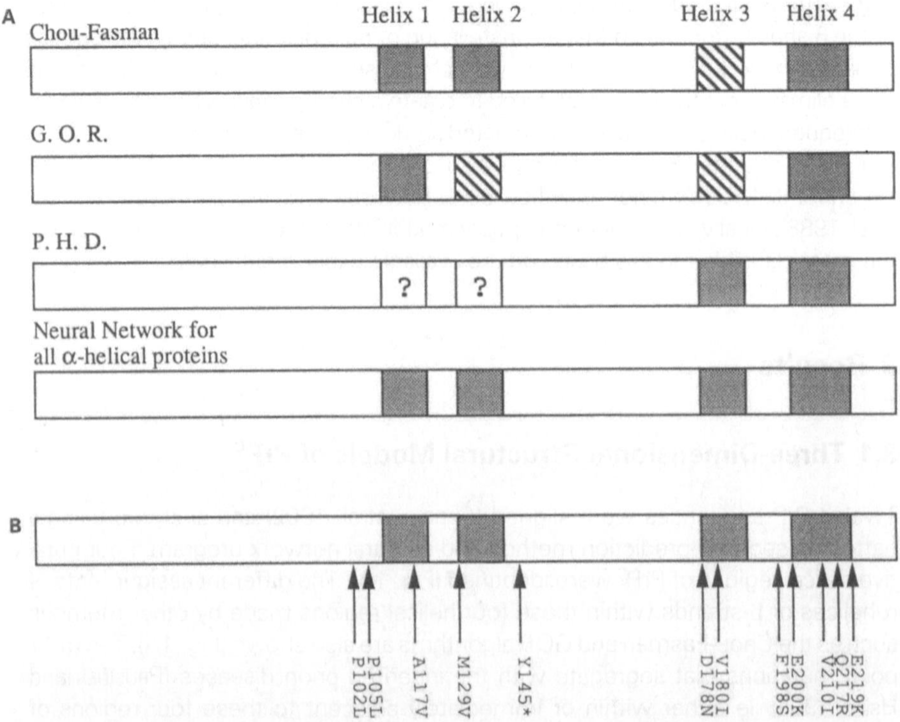

Fig. 1. Structural and mutational features of the PrP sequence. **A** Assignments of secondary structures within the four helical regions using the Chou-Fasman method (CHOU and FASMAN 1978), GOR algorithm (GARNIER et al. 1978), and PHD program (ROST and SANDER 1992). These four regions were identified as those likely to contain secondary structures using an approach of combining the pattern-based turn prediction (COHEN et al. 1986a) and polypeptide chain flexibility calculation (KARPLUS and SCHULZ 1985). The final assignment of α-helices within these regions was based on results from the neural network program optimized for all α-helical proteins (KNELLER et al. 1990; PRESNELL et al. 1993). *Stippled segments*, α-Helices; *diagonal lines*, β-sheets. **B** Point mutations that segregate with inherited prion diseases in humans. Four of these mutations have been genetically linked: P102L, D178N, F198S, and E200K (DLOUHY et al. 1992; GABIZON et al. 1993; HSIAO et al. 1989; PETERSEN et al. 1992). Polymorphisms M129V and E219K in human PrP gene are also shown (KITAMOTO and TATEISHI 1994; OWEN et al. 1990)

Table 1. Predicted α-helices and helix-helix interaction sites

Helix	Residue number[a]	Helix interaction sites[b]	Sequence identity (%)[c]
1	109–122	MKHM**A**G**A**AA**A**G**AV**V	100; 100; 100; 100; 100
2	129–141	MLGS**AM**SRPIIHF	92; 100
3	178–191	D**C**VNI**TI**KQHTVTT	100; 100; 58
4	202–218	DVKM**M**ER**VV**EQ**MC**ITQY	92; 100; 92; 100; 100

[a] The residue numbers correspond to those in human PrP. The corresponding numbers in the published alignment of the 12 PrP sequences are: H1(122–135), H2(142–154), H3(191–204) and H4(225–241) (GABRIEL et al. 1992).

[b] Human PrP amino acid sequences for the four predicted helices (1, 2, 3, and 4) as shown. Helix-helix interaction sites are highlighted by bold type and underlined. Single-letter code for amino acids is as follows: A, Ala; C, Cys; D, Asp; E, Glu; F, Phe; G, Glu; H, His; I, Ile; K, Lys; L, Leu; M, Met; N, Asn; P, Pro; V, Val; W, Trp; Y, Tyr.

[c] The percentage of sequence identity is shown in order corresponding to respective helix-helix interaction residues. The percentage is calculated as the number of identical residues divided by the number of total sequences.

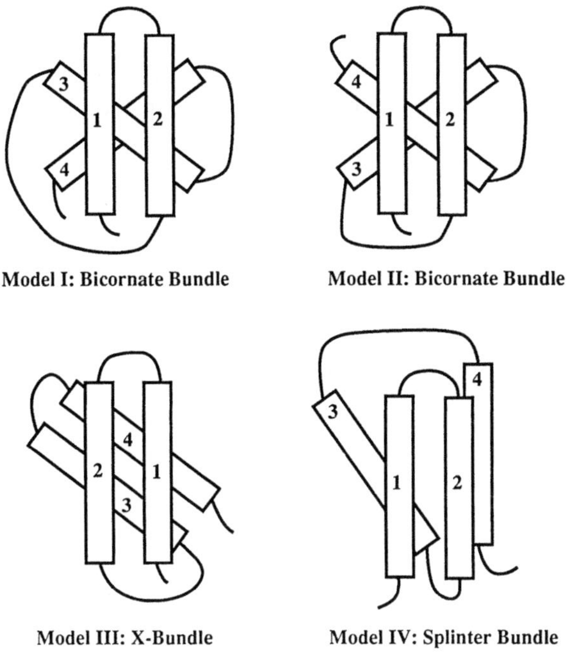

Fig. 2. Schematic drawings of the four plausible structures of PrP^C. These tertiary structural models were selected from approx. 300 000 possible structures based on the experimentally determined and theoretical constraints. The definitions for these models follow those used by HARRIS et al. (1994)

These four structures do not resemble the commonly observed square four-helix bundles studied by PRESNELL and COHEN (1989); however, they are compatible with a more diverse collection of four-helix bundles (HARRIS et al. 1994). HARRIS et al. (1994) found that helical bundles adopt a number of different topological arrangements in addition to the square bundle motif. According to their definitions, the four possible structures predicted for PrP^C are classified as bicornate, X-, and splinter bundles. Both models I and II belong to the bicornate type but display different relative locations for helices 3 and 4. Bicornate bundles have only been seen as part of a larger protein structure (HARRIS et al. 1994) while X- and splinter bundles seem to be stable in isolation without additional structural elements. For example, granulocyte-macrophage colony-stimulating factor is a splinter bundle (DIEDERICHS et al. 1991). The POU-specific DNA-binding domain of Oct-1 forms an X-bundle conformation (DEKKER et al. 1993).

Each of the four-helix bundle models was analyzed with respect to the point mutations found in the inherited prion diseases. The X-bundle structure was most intriguing with respect to the spatial clustering of mutations. The average distance between mutations on the different helices was significantly smaller for the X-bundle structure (9.3 Å) than the other three models (I, 14.2 Å; II, 14.5 Å; IV, 12.6 Å). The three-dimensional arrangement of the four putative helices in the X-bundle structure is shown in Fig. 3. The sites of helix-helix interactions on the hydrophobibic faces of each helix are highlighted in the ball and stick model (Fig. 3, left). The five point mutations (A117V, D178N, V1801, V2101, Q217R) cluster around a central hydrophobic core (Fig. 3, right) while an additional five mutations lie just adjacent to the helices (not shown). The sites of the mutations clustering around the hydrophobic core are shown in the ball and stick model as in the codon 129 (M→V) polymorphism which controls whether the codon 178 (D →N) mutation causing neurodegeneration presenting with insomnia or dementia (GOLDFARB et al. 1992); residue 129 lies at the beginning of helix 2.

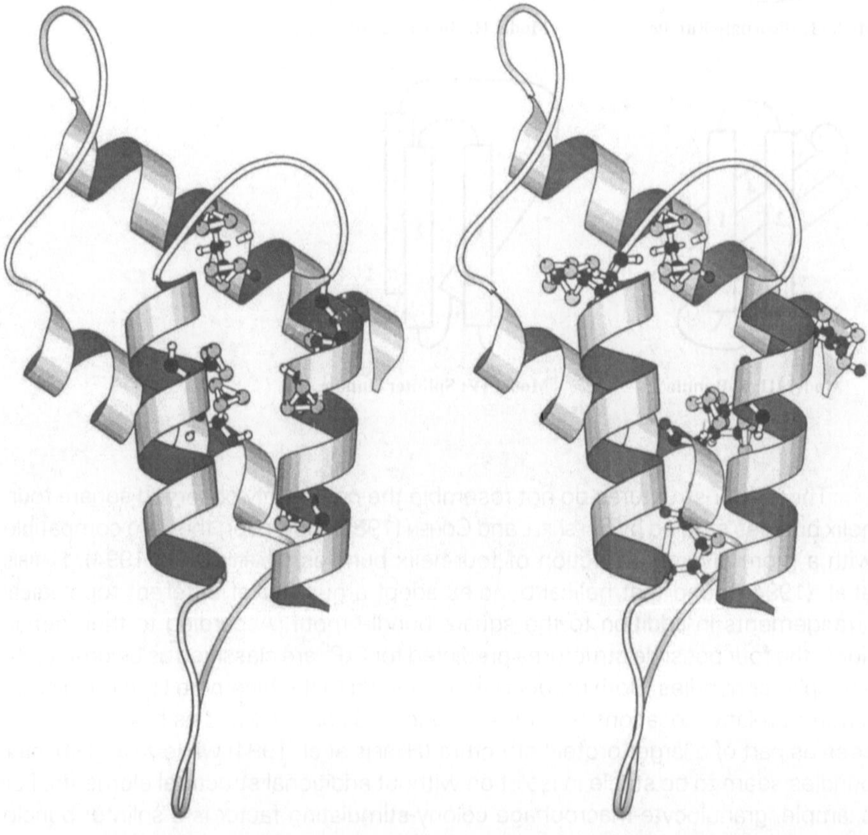

Fig. 3. The predicted three-dimensional structure of the X-bundle model of PrP^C. This structure was chosen from the four penultimate models of PrP^C (Fig. 2) because it seemed to correlate well with mutations that segregate with the inherited human prion diseases. The four helices 1, 2, 3, and 4 are displayed in the same orientation as that of the X-bundle model shown in Fig. 2. *Left*, the predicted helix-helix interaction sites; *right*, Positions of mutations within the α-helices that segregate with inherited prion diseases and the codon 129 polymorphism. Five mutations cluster around the central hydrophobic core of the proposed structure of PrP^C. We suggest that these mutations destabilize the structure of PrP^C, and thereby facilitate the conformational change that features in the formation of PrP^{Sc}

3.2 Three-Dimensional Structural Models of PrP^{Sc}

From an initial list of approx. 10^6 structures, six families of structures of PrP^{Sc} were considered viable (Fig. 4). These six structural models were validated with the profile method (LUTHY et al. 1992). The structural models of PrP^{Sc} were examined based on the available experimental data. Genetic experiments indicated that a specific interaction between PrP^C and PrP^{Sc} is important for prion propogation (PRUSINER et al. 1990; SCOTT et al. 1989). A small number of residues were found to be central to the prion species barrier, a phenomena that depends upon the homotypic interaction of PrP^C with PrP^{Sc} (SCHÄTZL et al. 1995). Each of these models was analyzed with respect to the residues implicated in PrP^C–PrP^{Sc} interactions. One of the models was most intriguing with respect to the spatial clustering of these residues. As shown in Fig. 5, this model displays a four-stranded β-sheet packed by two α-helices on the same side. The β-strand arrangement in the β-sheet is similar to that observed in the structure of the immunoglobulin binding domain of protein G (GB1) (GRONENBORN et al. 1991) while the overall packing topology between the β-sheet and two α-helices is similar to that observed in the structure of the RNA-binding domain of the human hnRNP C protein (WITTEKIND et al. 1992). Residues (Asn-108, Met-112, Met-129, and Ala-133) implicated in the species barrier, and thus PrP^{Sc}–PrP^C interactions are clustered on the solvent accessible surface of the β-sheet.

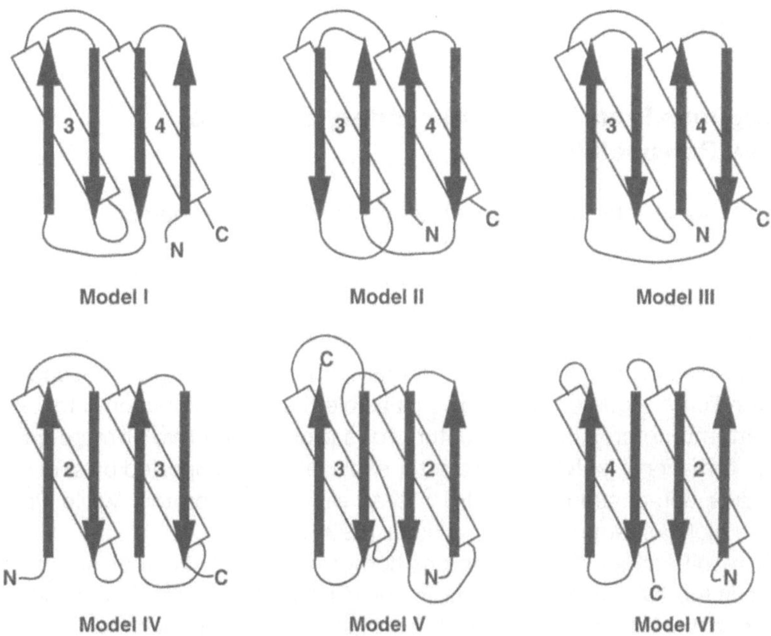

Fig. 4. Schematic drawings of the six plausible structures of PrP^{Sc}. These structural models were selected from an initial list of approx. 10^6 structures

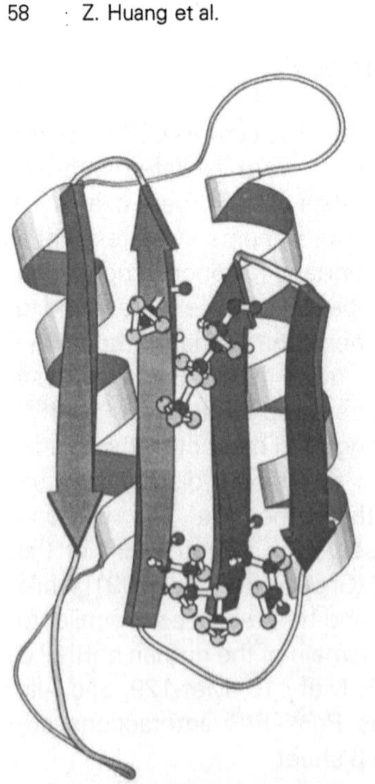

Fig. 5. The proposed three-dimensional structure of PrP^{Sc}. This structure was chosen from the six penultimate models of PrP^{Sc} (Fig. 4) because it seemed to correlate best with genetic data on residues involved in species barrier. It displays a four-stranded β-sheet on the front packed against by two α-helices on the back. Strands 1a and 1b correspond to the helix 1 in PrP^{C} (*dark gray arrows*); strands 2a and 2b correspond to the helix 2 (*light gray arrows*). Helices 3 and 4 in this model correspond to the same helices in PrP^{C} (HUANG et al. 1994). Four residues (Asn-108, Met-112, Met-129, and Ala-133) implicated in species barrier (SCHÄTZL et al. 1995) cluster onto the solvent accessible surface of the β-sheet

4 Discussion

4.1 Ambiguous Prediction of Secondary Structure: A New Perspective?

In our initial analysis of the secondary structure of PrP^{C}, general purpose methods such as the CHOU-FASMAN and GOR algorithms were used (CHOU and FASMAN 1978; GARNIER et al. 1978). These methods gave a confusing view of the secondary structural elements. In particular, these methods disagreed as to the preferred conformation for each of the four putative structural regions (Fig. 1A). While the conflicting predictions could reflect the intrinsic limitations of current secondary structure prediction algorithms, it is also plausible that apparent inconsistencies reflect an attempt to predict one structure for a protein with two or more quite distinct conformational isoforms. This latter explanation is supported by spectroscopic studies demonstrating that PrP^{C} is an all α-helical protein while PrP^{Sc} possesses a high β-sheet content (PAN et al. 1993).

These results suggest that different conformations of the four structural regions predicted by classical algorithms could be indicative of the alternative conformations that PrP can adopt. This formulation is consistent with experimental studies with synthetic peptides corresponding to the four putative regions of secondary structure for PrP. Three of the four peptides were found to adopt a β-sheet conformation in aqueous buffers (GASSET et al. 1992). This is remininscent of the conformations of the prepro PTH precursor peptide. The Chou-Fasman algorithm predicted two structures, one of which is largely α-helical and the other rich in β-strands (ROSENBLATT et al. 1980). CD studies demonstrated that this peptide can adopt two distinct conformations under different solvent conditions. Although it is widely accepted that some domains of proteins can adopt distinct conformations in different environments (MOTTONEN et al. 1992), structure prediction analyses generally attempt to identify a consensus conformation. While we realize the importance of seeking a consensus among various methods used for prediction of protein structure, we suggest that this conventional approach should probably be revised for proteins exhibiting conformational heterogeneity such as PrP. This was clearly necessary in understanding the conformational gymnastics of the influenza hemagglutinin protein (BAKER and AGARD 1994; BULLOUGH et al. 1994; CARR and KIM 1994). The results of studies on PrP summarized here provide a new perspective into the use and interpretation of secondary structure prediction methods.

4.2 Structural Implications of Mutations in PrP Genes

Since the first study demonstrating genetic linkage between a mutation in codon 102 and development of GSS Gerstmann-Sträussler-Scheinker syndrome (HSIAO et al. 1989), 17 other mutations in the PrP gene have been identified that segregate with the inherited prion diseases (PRUSINER and HSIAO 1994). In both humans and transgenic mice, these mutations produce neurodegeneration and often prions as demonstrated by transmission of disease to experimental animals. Presumably, these amino acid substitutions destabilize the conformation of PrP^{C} and facilitate its refolding into PrP^{Sc}. Ten of 11 known point mutations in the PrP open reading frame lie within or adjacent to the four putative α-helical domains (Fig. 1B). In the X-bundle structure (Fig. 3), the five point mutations (A117V, D178N, V1801, V2101, Q217R) cluster within the central hydrophobic core (Fig. 3, right) while an additional five mutations lie just adjacent to the helices (not shown). Furthermore, those five mutations occurring around the central hydrophobic core are at or near amino acid residues which are predicted to be essential for helix-helix interactions (Table 1, Fig. 3 right). Thus, many of these point mutations may not only destabilize the local α-helical conformation but also affect the long-range tertiary interactions between helices.

Fatal familial insomnia and familial CJD are genetically linked to the same mutation at codon 178 (D→N) while the codon 129 (M→V) polymorphism controls whether patients present with insomnia or dementia (GOLDFARB et al. 1992). Studies of PrP^{Sc} in the brains of patients dying of these inherited prion diseases suggest that the combination of mutation at codon 178 and polymorphism at codon 129 produces two different conformations of PrP^{Sc} which determine the distinctive clinical and pathological features of these two diseases (MONARI et al. 1994). Based on our three-dimensional models of PrP^{C}, mutation at codon 178 would eliminate the negative charge that interacts with the positive end of the helix dipole and destabilizes helix 3 while M →V polymorphism at codon 129 could affect the helix-helix interface between helices 1 and 2 that is centered on residue A133 (Fig. 3).

Another mutation at codon 210 (V →I) that produces familial CJD (POCCHIARI et al. 1993; RIPOLL et al. 1993) appears to be a site for helix-helix interactions on helix 4 (Table 1). The V →I mutation would disrupt the helix 4–helix 1 interaction and thus perturb the structure of PrP^C (Fig. 3). Our structural models of PrP^C seem to provide a consistent structural basis for observations on the inherited prion diseases.

4.3 Structure of PrP^C-PrP^{Sc} Complex: Implications for Species Barrier

The transmission of prions from one species to another is accompanied by a prolonged incubation time on the initial passage while subsequent passages in the recipient host are considerably shorter. This species barrier has been shown to be due to differences in PrP sequence (PRUSINER et al. 1990; SCOTT et al. 1989). The allotypic interactions between variant PrP^C and PrP^{Sc} molecules appear to influence incubation times with different strains of prions (CARLSON et al. 1994; DICKINSON et al. 1968). A small number of residues have been found to be central to the prion species barrier (SCHÄTZL et al. 1995). It is interesting to note that these residues cluster on the solvent accessible β-sheet surface of the proposed structural model of PrP^{Sc} (Fig. 5). In addition, a number of hydrophobic side chains are located on this surface. These observations suggest that the solvent accessible surface of the β-sheet provides a plausible interface for a specific interaction of PrP^{Sc} with PrP^C that promotes the conformational change of PrP^C to PrP^{Sc} (Fig. 6). Mutations at this interface would alter the incubation times associated with different strains of prions. This suggests that prion strains result from the changes in the conformation and stability of the PrP^C–PrP^{Sc} complex. Such a model may provide a framework for understanding the structural basis of prion species barrier.

4.4 Conformational Model for Prion Replication

The findings from our structural studies of PrP suggest a conformational model for prion replication that, in principle, can explain the pathogenesis of infectious, inherited, and sporadic forms of prion disease (Fig. 7). This model depends upon kinetic rather than thermodynamic control. PrP^C folds to a metastable but accessible conformation. Stochastic fluctuations in the structure of PrP^C would generate a partially unfolded, monomeric structure, PrP*, that is an intermediate in the formation of PrP^{Sc}. PrP* can revert to PrP^C, be degraded, or form PrP^{Sc}. Normally the concentration of PrP* is low while the activation energy barrier to PrP^{Sc}

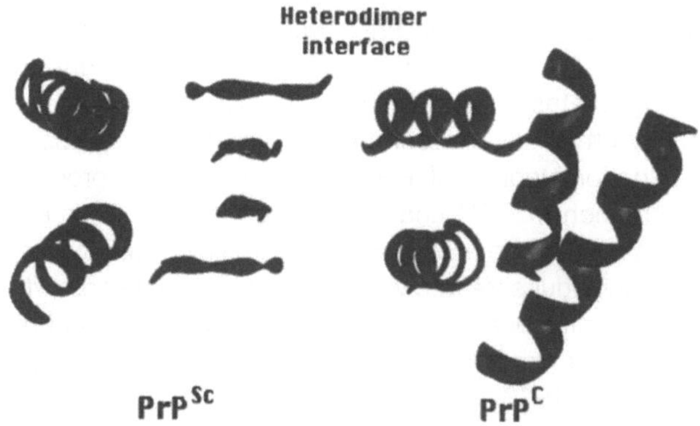

Fig. 6. A plausible structure for the PrP^{C}-PrP^{Sc} complex. In this heterodimeric complex the solvent accessible surface of the β-sheet of PrP^{Sc} packs against the hydrophobic surface of helices 1 and 2 of PrP^{C}. Residues (Asn-108, Met-112, Met-129, and Ala-133) involved in prion species barrier cluster onto the center of the PrP^{C}-PrP^{Sc} interface

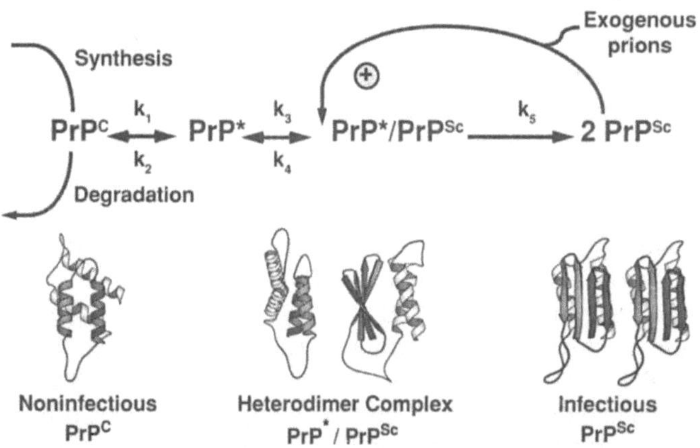

Fig. 7. A conformational model for prion replication. As shown, PrP^{C} is synthesized and degraded as part of the normal metabolism of many cells. Stochastic fluctuations in the structure of PrP^{C} can create (k_1) a rare partially unfolded, monomeric structure, PrP*, that is an intermediate in the formation of PrP^{Sc}, but can revert (k_2) to PrP^{C} or be degraded prior to its binding (k_3) into PrP^{Sc}. Normally, the concentration of PrP* is small and PrP^{Sc} formation is insignificant. In infectious prion diseases, exogenous prions enter the cell and stimulate conversion (k_5) of PrP* into PrP^{Sc} which is likely to be an irreversible process. In the case of inherited prion diseases, PrP mutations would destabilize PrP^{C} thereby lowering the barrier to the formation of PrP* (k_1) and subsequently PrP^{Sc} (k_5). Once this process has been initiated, template-assisted transformation of PrP* is expected. The sporadic prion diseases might result from the relatively infrequent accumulation of sufficient PrP* to produce PrP^{Sc}. Alternatively, somatic mutations might destabilize PrP^{C} which would promote its conversion into PrP*

formation is large. Therefore the concentration of PrP^{Sc} is insignificant. Since the insolubility of PrP^{Sc} has precluded analysis of its physical state, it remains uncertain whether PrP^{Sc} formation involves oligomerization. However, ionizing radiation data suggests that the minimal infectious particle is a dimer (Alper et al. 1967).

Infection with exogenous prions containing PrP^{Sc} would provide a template that promotes the conversion of PrP* into PrP^{Sc} by lowering the activation energy barrier that separates these two states. As shown in Fig. 7, using its β-sheet as a template, PrP^{Sc} could interact with PrP^{C} and direct the conformational process to generate new PrP^{Sc}. The superior stability of PrP^{Sc} would make this process irreversible and drive the formation of PrP* and thus additional PrP^{Sc} by mass action. Examples of prion infection include: kuru caused by ritualistic cannibalism among New Guinea, CJD in young adults treated with prion-contaminated growth hormone extracted from human pituitaries, and "mad cow" disease caused by prion-contaminated meat and bone meal prepared from the offal of scrapied sheep (Brown et al. 1992; Gajdusek 1977; Wilesmith 1992).

In the case of inherited prion diseases PrP mutations would destabilize PrP^{C} thereby lowering the barrier to the formation of PrP* and subsequently PrP^{Sc} (Fig. 7). Once the process has been initiated, template-assisted transformation of PrP* is expected. Current information on the location of PrP mutations (Dlouhy et al. 1992; Gabizon et al. 1993; Hsiao et al. 1989; Petersen et al. 1992; Poulter et al. 1992) is consistent with this hypothesis. As discussed above, 10 of 11 point mutations that segregate with inherited prion diseases fall within or near the four α-helices, and five mutations cluster within the central hydrophobic core that seems essential for the structural stability of PrP^{C} (Fig. 3).

The sporadic prion diseases might result from the relatively infrequent accumulation of sufficient PrP* to produce PrP^{Sc} (Fig. 7). In support of this proposal is the observation that transgenic mice overexpressing the wild-type PrP gene develop spongiform degeneration and produce prions in their brains (Westaway et al. 1994a). Alternatively, somatic mutations might destabilize PrP^{C}, which would promote its conversion into PrP*. Of note is the common Met/Val polymorphism at amino acid 129 in human PrP. Homozygosity at codon 129 appears to predispose to sporadic CJD (Palmer et al. 1991). Patients with the D178N mutation present with insomnia if codon 129 specifies Met in the mutant allele or dementia if codon 129 specifies Val (Goldfarb et al. 1992; Monari et al. 1994). Interestingly, the amino acid 129 lies at the β-sheet surface of PrP^{Sc} which seems to be an important interface for molecular recognition (Fig. 5). These structural findings may provide clues to the clinical observation described above.

Finally, based on the structures modeled for PrP^{C} and PrP^{Sc} and a variety of pieces of genetic, transgenic, and biophysical data, the following "infectious" prion cycle can be proposed (Fig. 8). An infectious dimer (or other multimer) of PrP^{Sc} binds an as yet unidentified chaperone protein X. This heterotrimeric complex mixes with PrP^{C} to form a heterotetramer. This complex facilitates the conversion of PrP^{C} to PrP^{Sc} and protein X can dissociate. The trimeric PrP^{Sc} complex can then bind another PrP^{C} molecule. This homotrimer can act as a template to facilitate the formation of another PrP^{Sc} molecule from PrP^{C}. Protein X can then facilitate the dissociation of the PrP^{Sc} homotetramer into two "infectious" PrP^{Sc} homodimers and form the heterotrimeric PrP^{Sc}/X complex completing the replication cycle.

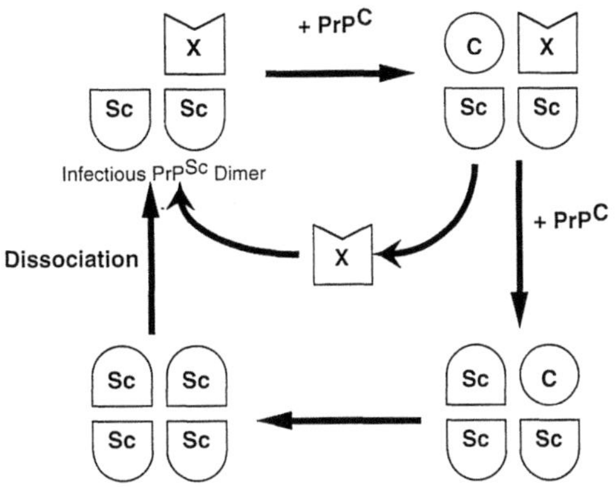

Fig. 8. A hypothetical scheme for the cycle of prion infection. An infectious dimer (or other multimer) of PrP^{Sc} binds an as yet unidentified chaperone protein X. This heterotrimeric complex mixes with PrP^{C} to form a heterotetramer. This complex facilitates the conversion of PrP^{C} to PrP^{Sc} and protein X can dissociate. The trimeric PrP^{Sc} complex can then bind another PrP^{C} molecule. This homotrimer can act as a template to facilitate the formation of another PrP^{Sc} molecule from PrP^{C}. Protein X can then facilitate the dissociation of the PrP^{Sc} homotetramer into two "infectious" PrP^{Sc} homodimers and form the heterotrimeric PrP^{Sc}/ X complex completing the replication cycle

4.5 New Approaches to Prion Diseases and Therapies

The predicted structure of PrP^{C} and PrP^{Sc} reported here provides a guide to further experimental studies. For example, by combining transgenetic approaches and computer modeling techniques, site-directed mutations at or near the helix-helix interaction residues (Table 1, Fig. 3) could disrupt or stabilize the structure of PrP^{C} and produce mice which are either susceptible or resistant to prion disease. The structure of PrP^{C}-PrP^{Sc} complex (Fig. 6) also provides an attractive target for transgenetic study. Mutations at the β-sheet surface of PrP^{Sc} could modulate its interaction with PrP^{C} and consequently the incubation time. This might offer a promising approach to the problem of species barriers or prion strains.

The structural model proposed for PrP^{C} and PrP^{Sc} might form the basis for the rational design of a molecule that prevents neurodegeneration. Since gene targeted mice with ablated PrP genes (Prn-$p^{0/0}$) are resistant to scrapie prions (Büeler et al. 1993; Prusiner et al. 1993), one rational therapy utilizes the antisense approach. However, there remain difficulties with delivering antisense polynucleotides to all neurons of the CNS. Perhaps, an approach more likely to succeed involves the design of some "molecular glue" that would cross the blood brain barrier and bind to the "hydrophobic core" of PrP^{C}. Such a pharmacotherapeutic would stabilize the surrounding α-helices and thus, might be effective in preventing PrP^{Sc} formation. An alternative approach is to design molecules that could bind to the interaction surface of infectious PrP^{Sc} and thus terminate its ability to interact and convert normal PrP^{C}. If such a nontoxic drug can be developed, it can be tested in mice with the sporadic, genetic, and infectious forms of prion disease (Hsiao et al. 1990; Westaway et al. 1994a) and subsequently in humans dying of these disorders.

Acknowledgments. Some of the material in this chapter is closely adopted from the work of Huang et al. (1994). We thank Drs. Peter A. Kollman, Irwin D. Kuntz, Robert J. Fletterick, Micheal Baldwin, Thomas L. James, Gerald D. Fasman and Bruno H. Zimm for insightful discussion. This work was supported by research grants from the National Institutes of Health and a gift from the Sherman Fairchild Foundation.

References

Alper T, Cramp WA, Haig DA, Clarke MC (1967) Does the agent of scrapie replicate without nucleic acid? Nature 214: 764–766

Baker D, Agard DA (1994) Influenza hemagglutinin: kinetic control of protein function. Structure 2: 907–910

Basler K, Oesch B, Scott M, Westaway D, Wälchli M, Groth DF, McKinley MP, Prusiner SB, Weissmann C (1986) Scrapie and cellular PrP isoforms are encoded by the same chromosomal gene. Cell 46: 417–428

Benner SA, Gerloff D (1991) Patterns of divergence in homologous proteins as indicators of secondary and tertiary structure: a prediction of the structure of the catalytic domain of protein kinases. Adv Enzyme Regul 31: 121–181

Boissel JP, Lee WR, Presnell SR, Cohen FE, Bunn HF (1993) Erythropoietin structure-function relationships – mutant proteins that test a model of tertiary structure. J Biol Chem 268: 15983–15993

Brown P, Preece MA, Will RG (1992) "Friendly fire" in medicine: hormones, homografts, and Creutzfeldt-Jakob disease. Lancet 340: 24–27

Büeler H, Aguzzi A, Sailer A, Greiner R-A, Autenried P, Aguet M, Weissmann C (1993) Mice devoid of PrP are resistant to scrapie. Cell 73: 1339–1347

Bullough PA, Hughson FM, Skehel JJ, Wiley DC (1994) Structure of influenza hemagglutinin at the PH of membrane fusion. Nature 371: 37–43

Carlson GA, Ebeling C, Yang S-L, Telling G, Torchia M, Groth D, Westaway D, DeArmond SJ, Prusiner SB (1994) Prion isolate specified allotypic interactions between the cellular and scrapie prion proteins in congenic and transgenic mice. Proc Natl Acad Sci USA 91: 5690–5694

Carr CM, Kim PS (1994) Flu virus invasion-halfway there. Science 266: 234–236

Caughey BW, Dong A, Bhat KS, Ernst D, Hayes SF, Caughey WS (1991) Secondary structure analysis of the scrapie-associated protein PrP 27–30 in water by infrared spectroscopy. Biochemistry 30: 7672–7680

Chothia C, Levitt M, Richardson D (1977) Structure of proteins: packing of alpha-helices and pleated sheets. Proc Natl Acad Sci USA 74: 4130–4134

Chou PY, Fasman GD (1978) Empirical predictions of protein conformation. Annu Rev Biochem 47: 251–276

Cohen FE, Richmond TJ, Richards FM (1979) Protein folding: evaluation of some simple rules for the assembly of helices into tertiary structures with myoglobin as an example. J Mol Biol 132: 275–288

Cohen FE, Sternberg MJE, Taylor WR (1982) The analysis and prediction of the tertiary structure and globular proteins involving the packing of α-helices against a β-sheet: a combinatorial approach. J Mol Biol 156: 821–862

Cohen FE, Abarbanel RM, Kuntz ID, Fletterick RJ (1986a) Turn prediction in proteins using a pattern-matching approach. Biochemistry 25: 266–275

Cohen FE, Kosen PA, Kuntz ID, Epstein LB, Ciardelli TL, Smith KA (1986b) Structure-activity studies of interleukin-2. Science 234: 349–352

Crick FHC (1953) The packing of α-helices: simple coiled coils. Acta Cryst 6: 689–697

Curtis BM, Presnell SR, Srinivasan S, Sassenfeld H, Klinke R, Jeffrey E, Cosman D, March CJ, Cohen FE (1991) Experimental and theoretic studies of the 3-dimensional structure of human interleukin-4. Proteins 11: 111–119

Dekker N, Cox M, Boelens R, Verrijer CP, Vandervliet PC, Kaptein R (1993) Solution structure of the POU-specific DN-binding domain of Oct-1. Nature 362: 852–855
Dickinson AG, Meikle VMH, Fraser H (1968) Identification of a gene which controls the incubation period of some strains of scrapie agent in mice. J Comp Pathol 78: 293–299
Diederichs K, Boone T, Karplus PA (1991) Novel fold and putative receptor binding site of granulocyte-macrophage colony-stimulating factor. Science 254: 1779–1782
Dlouhy SR, Hsiao K, Farlow MR, Foroud T, Conneally PM, Johnson P, Prusiner SB, Hodes ME, Ghetti B (1992) Linkage of the Indiana kindred of Gerstmann-Sträussler-Scheinker disease to the prion protein gene. Nat Genet 1: 64–67
Feng DF, Doolittle RF (1987) Progressive sequence alignment as a prerequisite to correct phylogenetic trees. J Mol Evol 25: 351–360
Gabizon R, Rosenmann H, Meiner Z, Kahana I, Kahana E, Shugart Y, Ott J, Prusiner SB (1993) Mutation and polymorphism of the prion protein gene in Libyan Jews with Creutzfeldt-Jakob disease. Am J Hum Genet 33: 828–835
Gabriel J-M, Oesch B, Kretzschmar H, Scott M, Prusiner SB (1992) Molecular cloning of a candidate chicken prion protein. Proc Natl Acad Sci USA 89: 9097–9101
Gajdusek DC (1977) Unconventional viruses and the origin and disappearance of kuru. Science 197: 943–960
Garnier J (1990) Protein structure prediction. Biochimie 72: 513–524
Garnier J, Osguthorpe DJ, Robson B (1978) Analysis of the accuracy and implications of simple methods for predicting the secondary structure of globular proteins. J Mol Biol 120: 97–120
Gasset M, Baldwin MA, Lloyd D, Gabriel J-M, Holtzman DM, Cohen F, Fletterick R, Prusiner SB (1992) Predicted α-helical regions of the prion protein when synthesized as peptides form amyloid. Proc Natl Acad Sci USA 89: 10940–10944
Gasset M, Baldwin MA, Fletterick RJ, Prusiner SB (1993) Perturbation of the secondary structure of the scrapie prion protein under conditions associated with changes in infectivity. Proc Natl Acad Sci USA 90: 1–5
Goldfarb LG, Petersen RB, Tabaton M, Brown P, LeBlanc AC, Montagna P, Cortelli P, Julien J, Vital C, Pendelbury WW, Haltia M, Wills PR, Hauw JJ, McKeever PE, Monari L, Schrank B, Swergold GD, Autilio-Gambetti L, Gajdusek DC, Lugaresi E, Gambetti P (1992) Fatal familial insomnia and familial Creutzfeldt-Jakob disease: disease phenotype determined by a DNA polymorphism. Science 258: 806–808
Gronenborn AM, Filpula DR, Essig NZ, Achari A, Whitlow M, Wingfield PT, Clore GM (1991) A novel, highly stable fold of the immunoglobulin binding domain of streptococcal protein G. Science 253: 657–661
Harris NL, Presnell SR, Cohen FE (1994) Four helix bundle diversity in globular proteins. J Mol Biol 236: 1356–1368
Hsiao K, Baker HF, Crow TJ, Poulter M, Owen F, Terwilliger JD, Westaway D, Ott J, Prusiner SB (1989) Linkage of a prion protein missense variant to Gerstmann-Sträussler syndrome. Nature 338: 342–345
Hsiao KK, Scott M, Foster D, Groth DF, DeArmond SJ, Prusiner SB (1990) Spontaneous neurodegeneration in transgenic mice with mutant prion protein. Science 250: 1587–1590
Huang Z, Gabriel J-M, Baldwin MA, Fletterick RJ, Prusiner SB, Cohen FE (1994) Proposed three-dimensional structure for the cellular prion protein. Proc Natl Acad Sci USA 91: 7139–7143
Hurle MR, Matthews CR, Cohen FE, Kuntz ID, Toumadje A, Johnson CW (1987) Prediction of the tertiary structure of the α-subunit of tryptophan synthase. Proteins 2: 210–224
Karplus PA, Schulz GE (1985) Prediction of chain flexibility in proteins. Naturwissenschaften 72: 212–213
Kitamoto T, Tateishi J (1994) Human prion diseases with variant prion protein. Philos Trans R Soc Lond [B] 343: 391–398
Kneller DG, Cohen FE, Langridge R (1990) Improvement in protein secondary structure prediction by an enhanced neural network. J Mol Biol 214: 171–182
Luthy R, Bowie JU, Eisenberg D (1992) Assessment of protein models with three-dimensional particles. Nature 356: 83–85
Marsh RF, Bessen RA, Lehmann S, Hartsough GR (1991) Epidemiological and experimental studies on a new incident of transmissible mink encephalopathy. J Gen Virol 72: 589–594
Medori R, Tritschler H-J, LeBlanc A, Villare F, Manetto V, Chen HY, Xue R, Leal S, Montagna P, Cortelli P, Tinuper P, Avoni P, Mochi M, Baruzzi A, Hauw JJ, Ott J, Lugaresi E, Autilio-Gambetti L, Gambetti P (1992) Fatal familial insomnia, a prion disease with a mutation at codon 178 of the prion protein gene. N Engl J Med 326: 444–449

Meyer RK, McKinley MP, Bowman KA, Braunfeld MB, Barry RA, Prusiner SB (1986) Separation and properties of cellular and scrapie prion proteins. Proc Natl Acad Sci USA 83: 2310–2314
Monari L, Chen SG, Brown P, Parchi P, Petersen RB, Mikol J, Gray F, Cortelli P, Montagna P, Ghetti B, Goldfarb LG, Gajdusek DC, Lugaresi E, Gambetti P, Autilio-Gambetti L (1994) Fatal fimilial insomnia and familial Creutzfeldt-Jakob disease: different prion proteins determined by a DNA polymorphism. Proc Natl Acad Sci USA 91: 2839–2842
Mottonen J, Strand A, Symersky J, Sweet RM, Danley DE, Geoghegan KF, Gerard RD, Goldsmith EJ (1992) Structural basis of latency in plasminogen activator inhibitor-1. Nature 355: 270–273
Oesch B, Westaway D, Wälchli M, McKinley MP, Kent SBH, Aebersold R, Barry RA, Tempst P, Teplow DB, Hood LE, Prusiner SB, Weissmann C (1985) A cellular gene encodes scrapie PrP 27–30 protein. Cell 40: 735–746
Owen F, Poulter M, Collinge J, Crow TJ (1990) Codon 129 changes in the prion protein gene in Caucasians. Am J Hum Genet 46: 1215–1216
Palmer MS, Dryden AJ, Hughes JT, Collinge J (1991) Homozygous prion protein genotype predisposes to sporadic Creutzfeldt-Jakob disease. Nature 352: 340–342
Pan K-M, Baldwin M, Nguyen J, Gasset M, Serban A, Groth D, Mehlhorn I, Huang Z, Fletterick RJ, Cohen FE, Prusiner SB (1993) Conversion of α-helices into β-sheets features in the formation of the scrapie prion proteins. Proc Natl Acad Sci USA 90: 10962–10966
Petersen RB, Tabaton M, Berg L, Schrank B, Torack RM, Leal S, Julien J, Vital C, Deleplanque B, Pendlebury WW, Drachman D, Smith TW, Martin JJ, Oda M, Montagna P, Ott J, Autilio-Gambetti L, Lugaresi E, Gambetti P (1992) Analysis of the prion protein gene in thalamic dementia. Neurology 42: 1859–1863
Pocchiari M, Salvatore M, Cutruzzola F, Genuardi M, Travaglini Allocatelli C, Masullo C, Macchi G, Alema G, Galgani S, Xi YG, Petraroli P, Silvestrini MC, Brunori M (1993) A new point mutation of the prion protein gene in familial and sporadic cases of Creutzfeldt-Jakob disease. Ann Neurol 34: 802–807
Poulter M, Baker HF, Frith CD, Leach M, Lofthouse R, Ridley RM, Shah T, Owen F, Collinge J, Brown G, Hardy J, Mullan MJ, Harding AE, Bennett C, Doshi R, Crow TJ (1992) Inherited prion disease with 144 base pair gene insertion. I. Genealogical and molecular studies. Brain 115: 675–685
Presnell SR, Cohen FE (1989) The topological distribution of four-α-helix bundles. Proc Natl Acad Sci USA 86: 6592–6596
Presnell SR, Cohen BI, Cohen FE (1993) MacMatch: a tool for pattern-based protein secondary structure prediction. Cabios 9: 373–374
Prusiner SB (1991) Molecular biology of prion diseases. Science 252: 1515–1522
Prusiner SB, Hsiao KK (1994) Human prion diseases. Ann Neurol 35: 385–395
Prusiner SB, McKinley MP, Bowman KA, Bolton DC, Bendheim PE, Groth DF, Glenner GG (1983) Scrapie prions aggregate to form amyloid-like birefringent rods. Cell 35: 349–358
Prusiner SB, Scott M, Foster D, Pan K-M, Groth D, Mirenda C, Torchia M, Yang S-L, Serban D, Carlson GA, Hoppe PC, Westaway D, DeArmond SJ (1990) Transgenetic studies implicate interactions between homologous PrP isoforms in scrapie prion replication. Cell 63: 673–686
Prusiner SB, Groth D, Serban A, Koehler R, Foster D, Torchia M, Burton D, Yang S-L, DeArmond SJ (1993) Ablation of the prion protein (PrP) gene in mice prevents scrapie and facilitates production of anti-PrP antibodies. Proc Natl Acad Sci USA 90: 10608–10612
Richmond TJ, Richards FM (1978) Packing of α-helices: geometrical constraints and contact areas. J Mol Biol 119: 537–555
Ring CS, Cohen FE (1993) Modeling protein structures-construction and their applications. FASEB J 7: 783–790
Ring CS, Cohen FE (1994) Conformational sampling of loop structure using genetic algorithms. Isr J Chem 34: 245–252
Ripoll L, Laplanche J-L, Salzmann M, Jouvet A, Planques B, Dussaucy M, Chatelain J, Beaudry P, Launay J-M (1993) A new point mutation in the prion protein gene at codon 210 in Creutzfeldt-Jakob disease. Neurology 43: 1934–1938
Rosenblatt M, Beaudette NV, Fasman GD (1980) Conformational studies of the synthetic precursor-specific region of preproparathyroid hormone. Proc Natl Acad Sci USA 77: 3983–3987
Rost B, Sander C (1992) Jury returns on structure prediction. Nature 360: 540
Safar J, Roller PP, Gajdusek DC, Gibbs CJ Jr (1993a) Conformational transitions, dissociation, and unfolding of scrapie amyloid (prion) protein. J Biol Chem 268: 20276–20284
Safar J, Roller PP, Gajdusek DC, Gibbs CJJ (1993b) Thermal-stability and conformational transitions of scrapie amyloid (prion) protein correlate with infectivity. Protein Sci 2: 2206–2216

Schätzl HM, Da Costa M, Taylor L, Cohen FE, Prusiner SB (1995) Prion protein gene variation among primates. J Mol Biol 245: 362–374
Scott M, Foster D, Mirenda C, Serban D, Coufal F, Wälchli M, Torchia M, Groth D, Carlson G, DeArmond SJ, Westaway D, Prusiner SB (1989) Transgenic mice expressing hamster prion protein produce species-specific scrapie infectivity and amyloid plaques. Cell 59: 847–857
Smith LJ, Redfield AC, Boyd AJ, Lawrence GM, Edwards R, Smith RA, Dobson CM (1992) Human interleukin 4. The solution structure of a four-helix bundle protein. J Mol Biol 224: 899–904
Stahl N, Baldwin MA, Teplow DB, Hood L, Gibson BW, Burlingame AL, Prusiner SB (1993) Structural analysis of the scrapie prion protein using mass spectrometry and amino acid sequencing. Biochemistry 32: 1991–2002
Turk E, Teplow DB, Hood LE, Prusiner SB (1988) Purification and properties of the cellular and scrapie hamster prion proteins. Eur J Biochem 176: 21–30
Westaway D, DeArmond SJ, Cayetano-Canlas J, Groth D, Foster D, Yang S-L, Torchia M, Carlson GA, Prusiner SB (1994a) Degeneration of skeletal muscle, peripheral nerves, and the central nervous system in transgenic mice overexpressing wild-type prion proteins. Cell 76: 117–129
Westaway D, Zuliani V, Cooper CM, Da Costa M, Neuman S, Jenny AL, Detwiler L, Prusiner SB (1994b) Homozygosity for prion protein alleles encoding glutamine-171 renders sheep susceptible to natural scrapie. Genes Dev 8: 959–969
Wilesmith J, Wells GAH (1991) Bovine spongiform encephalopathy. In: Chesebro BW (ed) Transmissible spongiform encephalopathics. Scrapie, BSE and related human disorders. Springer, Berlin Heidelberg New York, pp 21–38 (Current topics in microbiology and immunology, vol 172)
Wilesmith JW (1992) Bovine spongiform encephalopathy: a brief epidemiography, 1985–1991. In: Prusiner SB, Collinge J, Powell J, Anderton B (eds) Prion diseases of humans and animals. Ellis Horwood, London, pp 243–255
Wittekind M, Gorlach M, Friedrichs M, Dreyfuss G, Mueller L (1992) 1H, 13C, and 15N NMR assignments and global folding pattern of the RNA-binding domain of the human hnRNP C proteins. Biochemistry 31: 6254–6265

The Folding Intermediate Concept of Prion Protein Formation and Conformational Links to Infectivity

J. SAFAR

1 Introduction

Recent molecular biological and genetic evidence demonstrates that the abnormal isoform of the host cell derived prion protein plays a central role in the transmission and development of spongiform encephalopathies (GAJDUSEK 1988; PRUSINER 1993). The prion protein undergoes a three-stage transition during disease development: cellular isoform of prion protein (PrP^{C}) → scrapie isoform of prion protein (PrP^{Sc}) → prion protein (PrP27–30) (PRUSINER 1993).

Determining the molecular mechanism of prion protein transformation is critical to understand the transmission and pathogenesis of spongiform encephalopathies (prion diseases), as well as amyloid formation (GAJDUSEK 1988) and the resulting neuronal degeneration.

The PrP^{Sc} protein is an isoform of the host cell derived PrP^{C} protein (PRUSINER 1993). Both protein isoforms are encoded by the same cellular gene, and until now chemical differences between them have not been identified (for review see PRUSINER 1993). The secondary structure of PrP^{Sc} deduced from circular dichroism (CD) spectra (SAFAR et al. 1993a) has a higher β-sheet and lower α-helix content compared to PrP^{C} (PAN et al. 1993). It is possible that amyloid formation and disease transmission result solely from a conformational transition in the PrP monomer, and therefore it is essential to understand the initial step of this structural transformation. The resulting polymerization and proteolytic processing

Laboratory of Central Nervous System Studies, National Institute of Neurological Disorders and Stroke, National Institutes of Health, Bethesda, MD 20892, USA

into mature amyloid is a consequence of this first critical step, analogous to the unexplained seeding in nucleated polymerization hypotheses (COME et al. 1993; GAJDUSEK 1988). The transition is apparently triggered endogenously by point mutations or insertions in the PrP protein sequence, or exogenously by PrP^{Sc} or PrP27–30. The phenomenon of prion protein (scrapie amyloid) transformation adds an additional level of complexity to the general problem of determining the mechanism by which a specific amino acid sequence directs the folding into an unique conformation (JAENICKE 1991; MATTHEWS 1993).

The early stages and mechanisms of aggregation of abnormal isoforms of prion protein, as well as the secondary and tertiary structure, are very difficult to examine directly because of the prion protein's intrinsic insolubility and aggregated, polymer-like state. Additionally, scrapie and all other amyloids have monomeric conformational precursors with stable native secondary and tertiary structure. Thus, prion protein formation raises several questions not encountered in the folding and conformational transition of monomeric proteins: (a) Does the change in the secondary and tertiary structure of the prion protein monomer trigger aggregation, or is the conformational transition simply the result of aggregation? (b) Which conformational state(s) of PrP^{Sc} of PrP27–30 are able to replicate in vivo (be infectious), and what is the relationship of infectivity and prion protein (scrapie amyloid) structure? (c) If the native protein can form both normal monomers of PrP^{C} or abnormal aggregates of PrP^{Sc}, where in the folding pathway is the critical crossroad?

2 The Existence of General β-sheet (Amyloid) Conformational Pathway of Proteins in the Solid State

Amyloid formation may be described in the simplest terms as a transition from a monomeric solution into the solid state of aggregates, with the morphology of typical fibrils, amorphous aggregates, or films (SAFAR et al. 1993a). Because of the lack of data on solution → solid state protein conformational transitions, we first tested the possibility that aggregation by itself induces the β-sheet secondary structure. As model systems we used three synthetic peptides corresponding to the helical, turn or random protein structures, and three structurally well-described proteins: all-α helical myoglobin, mixed α-helical/β-sheet RNase A, and all-β-sheet structured concanavalin A. Using CD and infrared spectroscopy, together with high-resolution transmission electron microscopy, we directly compared the solution and aggregate conformations, and followed the stability of different secondary structure elements in transition from solution to aggregates in thin films (SAFAR et al. 1993c).

The experiments demonstrate that the solid state of aggregated proteins is noncrystalline and amorphous. Surprisingly, the transition from monomer in solution into the aggregated solid state induces no or only minor changes in the protein or peptide secondary structure. However, exposure of protein films to high temperature changes the various types of secondary structure, such as turns, α-helices, or random elements into uniform β-sheets. Once the films are β-sheet structured, they are highly stable even at temperatures above 100°C. The changes are irreversible, and after cooling to room temperature the proteins retain the new β-sheet secondary structure.

The data provide evidence that aggregation itself does not significantly change the secondary structure of model proteins or peptides. However, at higher temperatures, proteins and peptides in the solid state show a general conformational pathway with different elements of secondary structure converting into β-sheets (SAFAR et al. 1993c). Thus, any protein or peptide is potentially amyloidogenic and may be irreversibly converted into a β-sheet-structured solid state. However, in most proteins, such conformational transitions are separated by a large energy barrier, thereby preventing amyloid formation (SAFAR et al. 1993c).

3 Structure-Function Relationship of Prion Protein to Infectivity

The conformation of any protein is not static but controlled by the dynamic relationship of enthalpy and entropy of the protein's amino acid sequences and the solvent (SCHULZ and SCHIRMER 1979). As a result, various solvents may induce conformations different from that in water. When a protein is exposed to a different solvent in the solid state, it retains the new conformation even after eliminating the solvent in a vacuum (solvent "memory" effect; SAFAR et al. 1993c). This enables the secondary structure of the protein to change into a new and stable one. Because PrP27–30 reconstituted from dry films is infectious, we used this phenomenon to determine the impact of heat and organic solvents on the structure of PrP27–30 in the solid state and to correlate the resultant various stable conformations with the film's infectivity (SAFAR et al. 1993b).

The secondary structure analysis of the CD spectra of PrP27–30 dissociated in liposomes or aggregated in films is typical of all β-sheet-structured proteins (SAFAR et al. 1993a,b). However, aggregation and loss of free water in the solid state apparently increases the β-sheet content in the PrP27–30 secondary structure and alters the tertiary structure into a less ordered one (SAFAR et al. 1993b). β-Sheet growth in the aggregated state is probably due to an intermolecular alignment of external β-sheets and simultaneous intramolecular secondary structure transition favoring β-sheets. Such structural changes may simulate PrP27–30 conformational transitions in amyloid plaques following early aggregation of PrP27–30 in the extracellular space.

An important characteristic of agents causing spongiform encephalophathies (prion disease) is their exceptional thermal resistance (GAJDUSEK 1988). This thermoresistance is very difficult to explain, and in the case of the "protein only" prion model (PRUSINER 1993), it implies reversibility of the protein thermal unfolding or an absence of unfolding at high temperatures. Our experiments with purified PrP27–30 in thin films exposed to dry heat for 30 min indicate no change of infectivity at 100°C and only a minor drop at 132°C (SAFAR et al. 1993b). At the same time, the secondary structure of the PrP27–30 is heat-resistant in the solid state with no significant changes after exposure to 132°C (SAFAR et al. 1993b). The secondary structure of PrP27–30 in the solid state apparently has the necessary thermal stability to be a carrier of infectivity. The heat stability of the β-sheet-structured prion protein (scrapie amyloid) may be explained by the higher thermodynamic stability of the β-sheets and absence of free water in the solid state (SCHULZ and SCHIRMER 1979).

Solvent replacement studies of the PrP27–30 secondary structure in the solid state followed by CD spectroscopy indicate uniform conversion of native β-pleated-sheet-like secondary and tertiary structure into α-helices (SAFAR et al. 1993b). The strongest α-helix inducer was formic acid, a known scrapie infectivity inactivator (BROWN et al. 1990). The structural effect of formic acid may be explained by formation of temporary hydrogen bonds in the protein's secondary structure. During formic acid evaporation and in the absence of water, the formic acid hydrogen bonds are replaced by intrachain hydrogen bonds forming an α-helix in a zipperlike manner (SAFAR et al. 1993c). Simultaneously, the PrP27–30 film loses infectivity. Trifluoroacetic acid and fluorinated alcohols induce a different degree of β-to-α secondary and tertiary structure transitions of PrP27–30, and proportionally diminish the film's residual infectivity (SAFAR et al. 1993b).

These experiments provide evidence for a structure-function relationship between the native, β-pleated-sheet-like secondary and tertiary structure of prion protein, and infectivity. Surprisingly, inactivation of infectivity in the solid state is not associated with an anticipated denatured state but with transformation of β-sheets into α-helices (SAFAR et al. 1993b). Another possible interpretation is that the observed changes in prion protein conformation parallel the conformational changes of a still undetected subfraction of prion protein or scrapie agent responsible for scrapie transmission. However, it is possible that the β-pleated-sheet secondary and specific tertiary structure of the PrP protein is solely responsible for its ability to propagate and amplify. The alternative explanation is that the native conformation of PrP27–30 is critical for ligand binding, and only the entire complex can propagate, amplify, and form amyloid. This does not diminish the key role of the PrP27–30 protein structure, because any alterations in PrP27–30 conformation would result in changed ligand binding (JAENICKE 1991) and in different infectivity. The biological data on various "strains" of infectivity isolated from animals with an identical primary sequence of PrP protein may support the ligand concept (WEISSMANN 1991) or, alternatively, posttranslational modification (PRUSINER 1993). The least conventional explanation, the existence of multiple stable conformations of the same primary sequence under the same conditions able to specifically replicate itself from passage-to-passage, is difficult to reconcile with the current models of protein folding and thermodynamics (JAENICKE 1991; MATTHEWS 1993).

4 The Folding Intermediate (Framework) Concept of Prion Protein Formation

Advances in fluorescence, CD, and nuclear magnetic resonance spectroscopy, with the capability of monitoring the changes in protein conformation occurring in vitro in a millisecond time scale and during equilibrium unfolding, have dramatically affected our understanding of how proteins fold and unfold (JAENICKE 1991; MATTHEWS 1993). Current data available from a variety of proteins demonstrate that they do not fold in a single step but rather through distinct conformational intermediates with nonnative, nondenatured secondary and tertiary structures (vectorial "framework" concept, for review see JAENICKE 1991; MATTHEWS 1993). During the first step (Fig. 1) in less than 5 ms protein molecules in many different unfolded states collapse into more compact structures, forming a few discrete populations of molecules with similar elements of secondary structure and nonpolar surfaces. The next phase, lasting 5–1000 ms, involves formation of a single population of molecules with a native-like secondary structure but still consisting of the unstable tertiary structure (molten globule intermediate; MATTHEWS 1993). Finally, in a phase which may last hours, the concerted formation of many noncovalent interactions throughout the protein molecules establishes the native tertiary and secondary structure (MATTHEWS 1993). Kinetic and thermodynamic laws require PrP^{C} and PrP^{Sc} to undergo an analogous stepwise conformational transition in vivo.

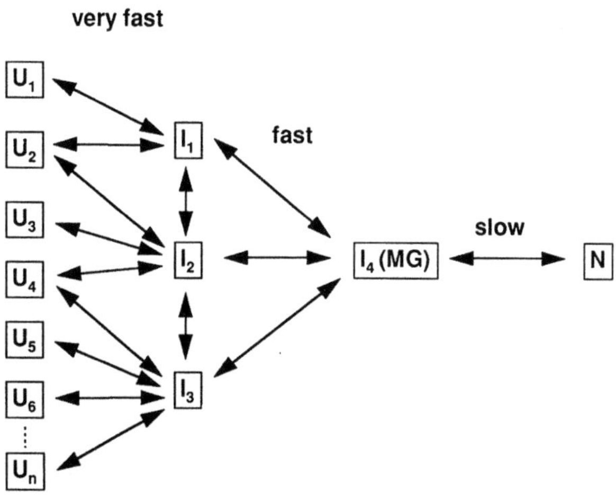

Fig. 1. The vectorial "framework" concept of protein folding pathways (JAENICKE 1991; MATTHEWS 1993). The protein initiates the folding process by undergoing a very fast transition (<5 ms) from a large number of unfolded states (U_1– U_n) to a limited subset of intermediates (I_1–I_3) with some elements of secondary structure; next is the much slower (5–1000 ms) transition from intermediates I_1–I_3 to a molten globule states (I_4, *MG*) with nativelike secondary structure and still unstable tertiary structure; and finally, the molten globule conformation slowly transforms into the native tertiary structure (*N*)

Therefore, the critical problems of scrapie amyloid (prion) protein formation are: (a) determining the intermediate steps in the folding pathway of PrP^C, (b) specifying the intermediate steps of the alternative folding pathway of PrP^{Sc}, and (c) explaining how both pathways are conformationally and thermodynamically linked to establish the critical crossroad (SAFAR et al. 1994).

Recent data from equilibrium dissociation and unfolding of PrP^{Sc}, and in more detail on its proteinase-resistant core PrP27–30, corroborate the framework concept of PrP protein folding, and show that both proteins fold through monomeric conformational intermediates (Fig. 2; SAFAR et al. 1993a, 1994). Such conformational intermediates of PrP27–30 are critical becauses (a) they are the direct thermodynamic and conformational precursors of the scrapie amyloid aggregate, and (b) comparision of the monomeric intermediates and PrP27–30 aggregate provides essential information about the stability, mechanism of association, and accompanying conformational transitions (MATTHEWS 1993; SAFAR et al. 1994).

The conformation of the prion protein folding intermediate, dissociated from aggregates at low concentrations of Gdn HCl or by decreasing pH, meets the definition of a molten globule folding intermediate (Fig. 2; SAFAR et al. 1993a, 1994). This state is structurally and thermodynamically distinct from both the native

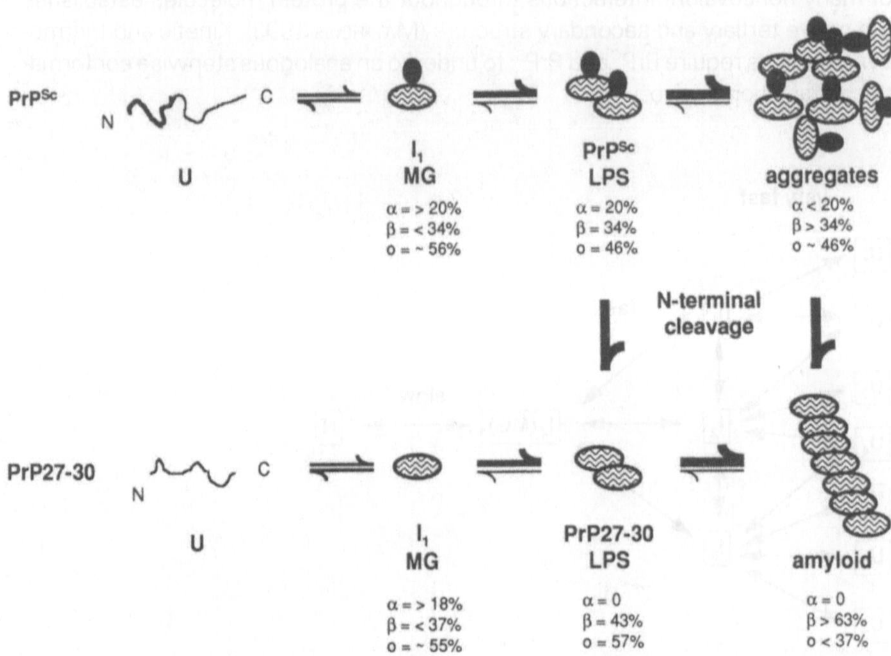

Fig. 2. The conformational pathway of prion protein (scrapie amyloid). The formation of PrP^{Sc} and PrP27–30 through unfolded (*U*), molten globule (*MG*), liposome-dissociated (*LPS*), and aggregated states is based on the data obtained by equilibrium dissociation and unfolding (SAFAR et al. 1993a, 1994), and by the secondary structure content analysis of CD spectra. The data indicate a continuing increase in β-sheet (β) secondary structure content from monomeric molten globule to the highest in aggregates at the expense of α-helical (α), random and turn (o) secondary structure elements. The existence of dimers of PrP^{Sc}/PrP27–30 in liposomes is hypothetical

and denatured states of the protein. It is defined as a compact, partly ordered conformation with nonspecific tertiary structure, high secondary structure content, significant structural flexibility, and an intrinsic tendency to aggregate as a result of a more exposed hydrophobic surface (KUWAJIMA 1989; MATTHEWS 1993).

The molten globule state of any protein is significantly less stable than its native form (KUWAJIMA 1989). This lower stability of molten globule conformation could provide a thermodynamic "tunnel" for the transition from a largely α-helical PrP^{C} into more β-structured PrP^{Sc} and aggregates. This phenomenon may be an important mechanism to facilitate $PrP^{C} \rightarrow PrP^{Sc}$ conversion because such a process is separated by a large energy barrier in model proteins and peptides (SAFAR et al. 1993c). The hypothetical ligand or binding with chaperonins could stabilize PrP^{Sc} in the molten globule form thereby further favorably diminishing the energy barrier.

The complexity and more than two-state equilibrium dissociation and unfolding of PrP^{Sc} and PrP27–30 preclude precise calculation of the stability (ΔG^{app}) of the proteins (SAFAR et al. 1993a, 1994). However, the midpoints of the conformational transitions of each intermediate are proportionate to their stability. By analyzing the transition midpoints it is apparent that the PrP^{Sc} or PrP27–30 aggregates are significantly more stable than the molten globule conformation (SAFAR et al. 1993a) and therefore may represent a thermodynamic trap for normally monomeric and metastable conformational intermediates (Fig. 3). Additional stabilization with a

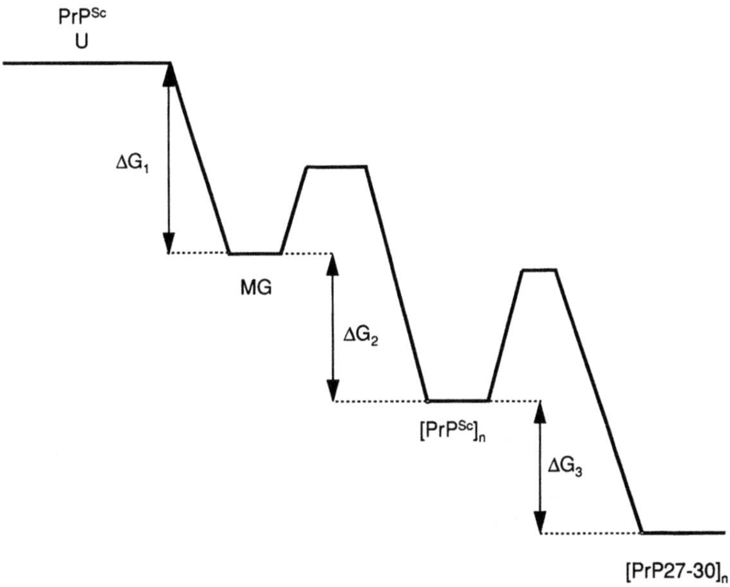

Fig. 3. The formal thermodynamic description of PrP^{Sc} and PrP27–30 conformational pathways. The pathways of PrP^{Sc} and PrP27–30 with unfolded (*U*) conformation, molten globule (*MG*) conformation, and aggregates of $[PrP^{Sc}]_n$ and $[PrP27\text{–}30]_n$ were determined by equilibrium dissociation and unfolding of both proteins in Gdn HCl (SAFAR et al. 1993a, 1994). The molten globule intermediate is a critical initial step triggering aggregation, and aggregates of $[PrP^{Sc}]_n$ may be understood as a thermodynamic trap for this otherwise intrinsically unstable protein conformation. The plotted apparent energy of transition from one state to another ($\Delta G_1 - \Delta G_4$) and the activation energy barrier between them are arbitrary

concurrent increase in β-sheet content in the secondary structure is achieved by proteolytic cleavage of the N-terminal regions of PrP^{Sc} and realignment of PrP27–30 into mature amyloid (Figs. 2, 3; SAFAR et al.). Amino terminal cleavage is an important additional irreversible step with the resultant aggregates being more stable than those of PrP^{Sc} (SAFAR et al. 1993a). Thermodynamic stability of the protein is closely related to the turnover rate (JAENICKE 1991), and the high stability of PrP^{Sc} and PrP27–30 aggregates alone may account for their accumulation.

The conformations, defined as molten globule intermediates, exist either as kinetic protein folding intermediates (MATTHEWS 1993) or equilibrium intermediates (KUWAJIMA 1989). The compact early kinetic intermediates of protein folding pathways occur during the early folding phases (MATTHEWS 1993), and are probably the universal step in protein folding. The conformationally similar metastable protein forms appear in equilibrium in some apoproteins, or are stabilized at equilibrium by low pH (A forms) or by interaction with chaperonins (SAFAR et al. 1994). Which mechanisms are involved in the formation of the monomeric molten globule intermediate of prion protein and result in aggregation and amyloid formation in vivo is the critical question in the pathogenesis of the disease.

Acknowledgments. The authors thank Devera G. Schoenberg for editing the manuscript, and Steve Ono for document production assistance.

References

Brown P, Wolff A, Gajdusek DC (1990) A simple and effective method for inactivating virus infectivity in formalin-fixed tissue samples from patients with Creutzfeldt-Jakob disease. Neurology 40: 887–890

Come JH, Fraser PE, Lansbury PT (1993) A kinetic model for amyloid formation in the prion diseases-importance of seeding. Proc Natl Acad Sci USA 90: 5959–5963

Gajdusek DC (1988) Transmissible and nontransmissible dementias: distinction between primary cause and pathogenetic mechanism in Alzheimer's disease and aging. Mt Sinai J Med 55: 3–5

Jaenicke R (1991) Protein folding: local structures, domains, subunits, and assemblies. Biochemistry 30: 3147–3161

Kuwajima K (1989) The molten globule state as a clue for understanding the folding and cooperativity of globular-protein structure. Proteins Struct Funct Genet 6: 87–103

Matthews CR (1993) Pathways of protein folding. Annu Rev Biochem 62: 653–683

Pan K-M, Baldwin M, Nguyen J, Gasset M, Serban A, Groth D, Mehlhorn I, Huang Z, Fletterick RJ, Cohen FE, Prusiner SB (1993) Conversion of α-helices into β-sheets features in the formation of scrapie prion proteins. Proc Natl Acad Sci USA 90: 10962–10966

Prusiner SB (1993) Genetic and infectious prion diseases. Arch Neurol 50: 1129–1153

Safar J, Roller PP, Gajdusek DC, Gibbs CJ Jr (1993a) Conformational transitions, dissociation, and unfolding of scrapie amyloid (prion) protein. J Biol Chem 268: 20276–20284

Safar J, Roller PP, Gajdusek DC, Gibbs CJ Jr (1993b) Thermal stability and conformational transitions of scrapie amyloid (prion) protein correlate with infectivity. Protein Sci : 2206–2216

Safar J, Roller PP, Ruben GC, Gajdusek DC, Gibbs CJ Jr (1993c) Secondary structure of proteins associated in thin films. Biopolymers 33: 1461–1476

Safar J, Roller PP, Gajdusek DC, Gibbs CJ Jr (1994) Scrapie amyloid (prion) protein has the conformational characteristics of an aggregated molten globule folding intermediate. Biochemistry 33: 8375–8383

Schulz GE, Schirmer RH (1979) Principles of protein structure. Springer, Berlin Heidelberg, New York, (Springer Advanced Texts in Chemistry)

Weissmann C (1991) A 'unified theory' of prion propagation. Nature 352: 679–683

Cell Biology of the Prion Protein

D.A. Harris, A. Gorodinsky, S. Lehmann, K. Moulder, and S.L. Shyng

1 Introduction

Prions are the infectious particles responsible for a group of fatal neurodegenerative diseases known as spongiform encephalopathies (reviewed in Prusiner and DeArmond 1994). This group includes Creutzfeldt-Jakob disease, kuru, Gerstmann-Sträussler syndrome, and fatal familial insomnia in human beings, and scrapie and bovine spongiform encephalopathy in animals. The principal component of prions is the glycoprotein PrP^{Sc}, which is a posttranslationally modified isoform of a normal cell-surface protein of the host called PrP^C (Oesch et al. 1985). PrP^C and PrP^{Sc} differ from each other in secondary structure and display markedly different biochemical properties (Caughey et al. 1991; Pan et al. 1993; Safar et al. 1993). Prion replication is hypothesized to occur when PrP^{Sc} in the infecting inoculum interacts specifically with host PrP^C, catalyzing its conversion to PrP^c (Prusiner et al. 1990). It is postulated that this conversion takes place spontaneously in PrP^{Sc} molecules carrying mutations that have been linked to familial forms of prion disease.

We have been interested in the cell biological properties of both PrP^{Sc} and PrP^C, in particular how these proteins are posttranslationally processed, and how

Department of Cell Biology and Physiology, Washington University School of Medicine, 660 South Euclid Avenue, St. Louis, MO 63110, USA

they are targeted to specific organelles within the cell. Our studies have had two major objectives. One has been to understand the cellular mechanisms underlying conversion of PrP^{C} into PrP^{Sc}. Recent studies have implicated both the endocytic and secretory pathways in this process. PrP^{C} is found on the plasma membrane, where it is attached by a glycosyl-phosphatidylinosital (GPI) anchor, while PrP^{Sc} resides inside the cell, localized in part to acid phosphatase-positive vesicles that are presumed to be lysosomes (TARABOULOS et al. 1990a; MCKINLEY et al. 1991). It has been reported recently that removal of PrP^{C} from the surface of prion-infected neuroblastoma cells using extracellular phospholipase or proteases blocks the appearance of PrP^{Sc} intracellularly (CAUGHEY and RAYMOND 1991; BORCHELT et al. 1992). This result is consistent with the idea that PrP^{C} on the cell surface is an essential precursor of PrP^{Sc} and suggests that conversion of the cellular to the scrapie isoform occurs either on the plasma membrane or intracellularly following endocytosis of PrP^{C}. Studies of glycosylation- and GPI-defective forms of PrP indicate that production of the infectious isoform might also occur along the secretory pathway (TARABOULOS et al. 1990b; ROGERS et al. 1993). Whether PrP^{Sc} can be generated in multiple compartments of the cell, and what their identity is, remain unknown.

The second objective of our studies has been to gain some insight into the physiological function of PrP^{C}, which has remained elusive. The protein is likely to be important, since it is abundantly expressed in the neurons of the CNS, and at lower levels in some peripheral tissues, beginning early in embryonic development (MANSON et al. 1992; HARRIS et al. 1993a), and since portions of its sequence have been highly conserved during evolution (GABRIEL et al. 1992). The presence of PrP^{C} on the cell surface argues for a role in adhesion, transmembrane signaling, or uptake of extracellular ligands. Mice in which the PrP gene has been deleted display abnormalities in GABA-mediated inhibition and long-term potentiation in the hippocampus, but whether PrP^{C} is directly involved in the synaptic transmission process is unclear (COLLINGE et al. 1994). We believe that analysis of the cellular processing and targeting of PrP^{C} are likely to provide additional clues to the physiological function of this isoform.

2 Posttranslational Processing of chPrP

Many of our initial studies focused on chPrP, the chicken homologue of mammalian PrP^{C}. ChPrP and mammalian PrP^{C}s are 30%–35% identical at the amino acid level and display the same set of structural domains. An uninterrupted stretch of 24 amino acids in the central portion of the molecule is identical in chPrP and mouse PrP. We have now confirmed with mouse and hamster PrP^{C}s most of the key findings with chPrP to be described below, and it is therefore likely that the essential cellular and biochemical properties of the avian and mammalian proteins are identical (LEHMANN et al. 1994).

We have used sequence-specific antibodies and protein chemical methods to characterize chPrP molecules in stably transfected cultures of N2a mouse neuroblastoma cells, and in chicken brain and cerebrospinal fluid (HARRIS et al. 1993b). Immunoblots of medium conditioned by chPrP-expressing N2a cells reveals two fragments of 30 and 10 kDa, which are detected with an antibody directed against the N-terminal half of the chPrP molecule (Fig. 1B, left). Fragments of 30 and 20 kDa are seen on immunoblots of cell lysates developed with a C-terminally directed antibody (Fig. 1B, right).

These results suggest that the chPrP molecule undergoes at least two cleavages as part of its normal metabolism. One (labeled "A" in Fig. 1A) occurs within the GPI anchor that attaches the protein to the plasma membrane. Since full-length protein released into the medium can be metabolically labeled with [^{3}H]- ethanolamine, cleavage A is likely to occur between the diacylglycerol moiety and the ethanolamine residue that is attached to the C-terminal amino acid of the protein. A second cleavage (labeled "B") is located within the region of 24 amino acids that is identical in chicken and mouse PrP. Edman degradation and mass spectroscopic analysis of the purified 10-kDa N-terminal fragment indicate that a major cleavage site is phenylalanine 116.

Over half of the chPrP molecules on the cell surface at steady-state are cleaved at site B, suggesting that this processing event may be important for the physiological function of the protein. One possibility is that the soluble N-terminal fragments serve as biologically active ligands that bind to a target cell. A number of polypeptide growth factors are derived by proteolytic cleavage of a membrane-anchored precursor (MASSAGUÉ 1990).

3 Endocytosis of chPrP

Several lines of evidence indicate that chPrP cycles constitutively between the cell surface and an endocytic compartment in transfected N2a cells (SHYNG et al. 1993). First, we have found that cleavage B (Fig. 1A) is inhibited by lysosomotropic amines such as ammonium chloride and by serine protease inhibitors such as leupeptin. This suggests that the cleavage occurs within an intracellular compartment that is both acidic and protease-containing, properties characteristic of endosomes. In addition, pulse-chase experiments and the use of the inhibitor brefeldin A indicate that the 10- and 20-kDa fragments resulting from cleavage B are not detectable until after newly synthesized chPrP molecules reach the cell surface, consistent the cleavage occurring along the endocytic rather than the secretory pathway.

Endocytosis of chPrP can be demonstrated directly using two different methods. First, internalization of antibodies bound to chPrP on the cell surface can be visualized using immunofluorescence microscopy (SHYNG et al. 1993). Second, internalization of surface-iodinated chPrP molecules can be measured

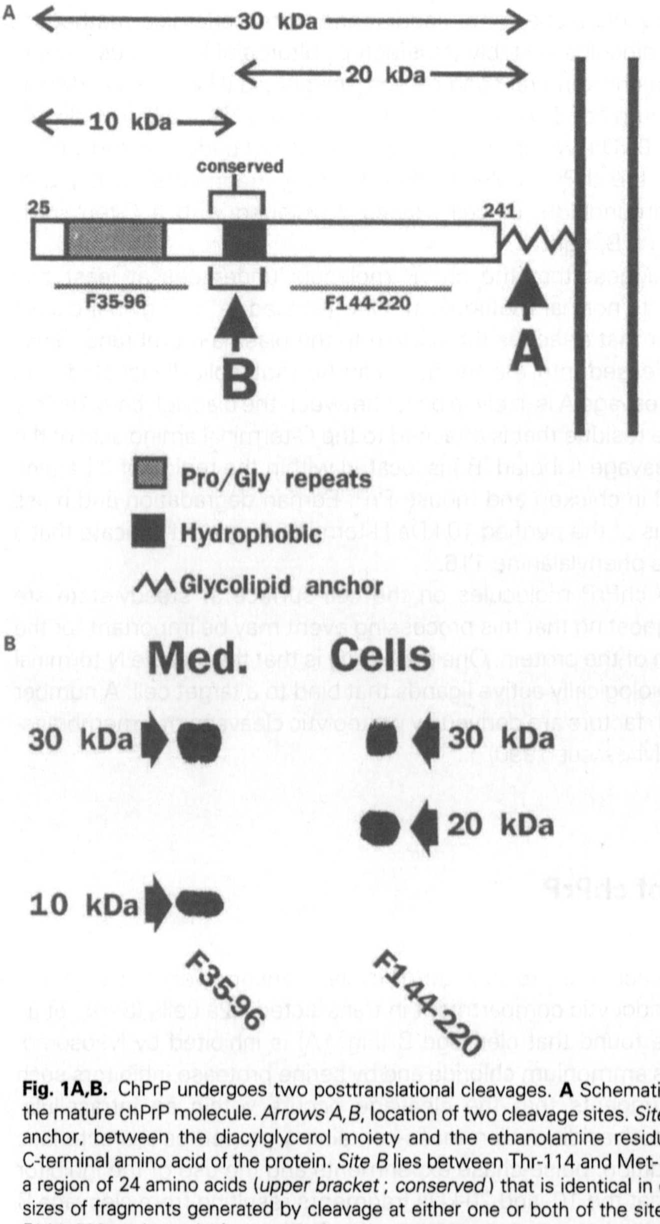

Fig. 1A,B. ChPrP undergoes two posttranslational cleavages. **A** Schematic showing the structure of the mature chPrP molecule. *Arrows A,B*, location of two cleavage sites. *Site A* lies within the glycolipid anchor, between the diacylglycerol moiety and the ethanolamine residue that is attached to the C-terminal amino acid of the protein. *Site B* lies between Thr-114 and Met-137 (*lower bracket*), within a region of 24 amino acids (*upper bracket*; *conserved*) that is identical in chPrP and mouse PrP. The sizes of fragments generated by cleavage at either one or both of the sites is indicated. F35–96 and F144–220 are bacterial fusion proteins encompassing the indicated segments of the chPrP sequence; antisera raised against these fusion proteins were used to develop the immunoblots in **B. B** Immunoblots of conditioned medium (*left*) and cell lysates (*right*) derived from chPrP-expressing N2a cells. The left-hand blot was developed with an antiserum to fusion protein F35–96, and the right-hand blot with an antiserum to fusion protein F144–220. The fragments (*arrows*) correspond to those illustrated on the schematic of the chPrP molecule in **A**. Samples were enzymatically deglycosylated before immunoblotting to facilitate comparison of the polypeptide chain lengths

by immunoprecipitation, after treatment of cells with phosphatidylinositol-specific phospholipase C (PIPLC) to separate surface from intracellular chPrP (Fig. 2). In this kind of experiment the distribution of chPrP approaches a steady state, with about half of the protein on the outside of the cell and half inside. This existence of a steady state suggests that internalized chPrP molecules are recycled back to the cell surface, a pathway that can be directly demonstrated by a variation of the surface iodination protocol (SHYNG et al. 1993).

Measurement of the rates of internalization and recycling from iodination experiments indicates that chPrP molecules cycle through N2a cells with a transit time of approx. 60 min. The rate of cleavage at site B is relatively slow, so that during each passage through the cell only 1%–5% of the molecules are clipped, and both intact and N-terminally truncated molecules are returned to the

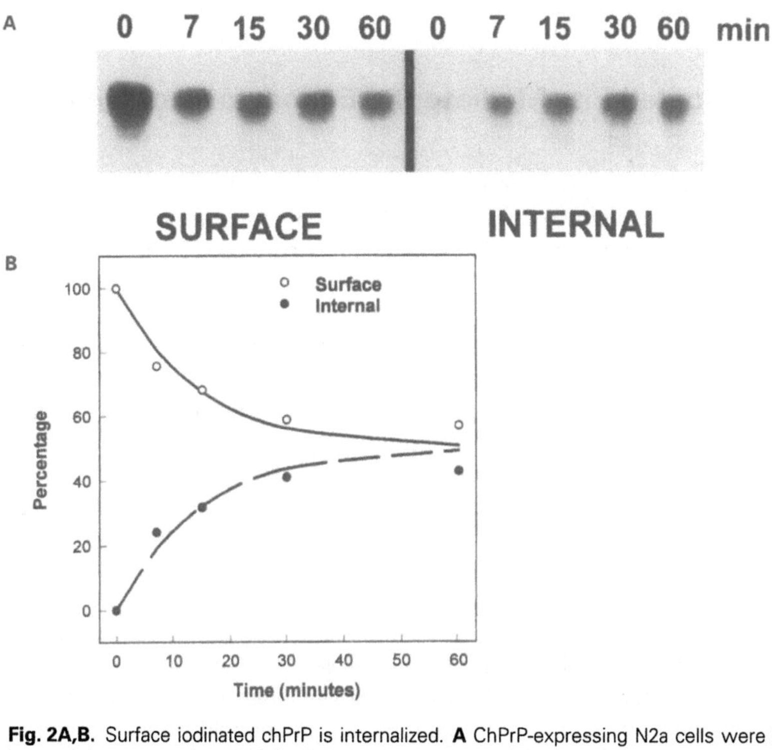

Fig. 2A,B. Surface iodinated chPrP is internalized. **A** ChPrP-expressing N2a cells were iodinated on ice, and then warmed to 37°C for 0, 7, 15, 30, or 60 min. Prior to lysis cells were treated with PIPLC for 2 h at 4°C to distinguish surface from internal chPrP. Samples were immunoprecipitated with antiserum F35–96, which recognizes the full-length form of chPrP shown here (35–45 kDa without deglycosylation). **B** The amounts of surface and internal chPrP at each time point were quantitated by densitometric scanning of autoradiograms. *Each point*, average value from four dishes. Data are expressed as a percentage of the total amount of chPrP immunoprecipitated from each dish and were corrected for PIPLC digestion efficiency (83% ± 4% for plates held at 4°C). Curves were fit by least-squares analysis to the equations $0.5\ [1 - e^{-kt}]$ (internal), or $0.5\ [1 + e^{-kt}]$ (surface). These equations assume a steady-state value of 1 for the ratio of internal/surface chPrP, and give a calculated $t_{½}$ of 10 min for the approach to steady state. (From SHUNG et al. 1993, with permission)

cell surface. Our results indicate that internalized chPrP is probably restricted to early compartments along the endocytic pathway, since little of the protein colocalizes with the lysosomal marker protein lgp120 by immunofluorescence microscopy.

The endocytic pathway we have demonstrated suggests the possibility that chPrP could serve as a receptor for an unknown extracellular ligand. The ligand might be a nutrient molecule, by analogy to transferrin and low-density lipoprotein, whose receptors constitutively cycle between the cell surface and early endosomes. Alternatively, chPrP might be a receptor for a growth factor or hormone, binding of which induces receptor internalization and degradation. Indeed, we have evidence that alterations in growth conditions can dramatically alter the amount of chPrP on the surface of neuroblastoma cells (unpublished data).

4 Clathrin-Coated Pits Mediate the Endocytosis of chPrP

Endocytic targeting of glycolipid-anchored proteins such as chPrP and other PrP^{C}s poses an interesting cell biological problem. Transmembrane receptors such as those for transferrin, low-density lipoprotein, and mannose 6-phosphate are internalized via clathrin-coated pits, a process that depends on a tyrosine-containing motif in the cytoplasmic tail of the receptor (reviewed by Trowbridge et al. 1993). GPI-anchored molecules lack a cytoplasmic domain and therefore cannot interact directly with intracellular adaptor molecules that are components of the clathrin coat. It has been reported that several glycolipid-anchored proteins are excluded from coated pits, although what alternative endocytic pathways are operative is unclear (Bretscher et al. 1980; Keller et al. 1992). Non-clathrin-coated invaginations of the plasma membrane called caveolae are seen in several cell types, and these structures have been proposed to mediate internalization of some glycolipid-anchored proteins, such as the receptor for the vitamin folic acid (Rothberg et al. 1990; Anderson 1993).

Our own results have led to the unexpected conclusion that chPrP is endocytosed via clathrin-coated pits, rather than caveolae, in N2a cells as well as in neurons and glia cultured from embryonic chicken brain (Shyng et al. 1994). Several lines of evidence support this idea. First, immunogold labeling of these cells demonstrates that clusters of chPrP molecules are closely associated with coated pits (Fig. 3). Morphometric quantitation indicates that the concentration of chPrP within 0.05 µm of coated pits is three to six times higher than over other areas of the plasma membrane. Moreover, gold particles can be seen within coated vesicles that are in the process of pinching off from the plasma membrane. Double labeling with antibodies to chPrP and to the transferrin receptor, a transmembrane protein known to be concentrated in coated pits, shows colocalization of the two proteins. In addition, internalization of chPrP is reduced by 70%

after incubation of N2a cells in hypertonic sucrose, a treatment that inhibits endocytosis by disrupting clathrin lattices (HEUSER and ANDERSON 1989). Finally, chPrP is present by immunoblotting in clathrin-coated vesicles purified from chicken brain.

Caveolae do not seem to be involved in the endocytosis of chPrP in N2a cells since we could not detect these structures by thin-section or deep-etch electron microscopy. In deep-etch images of fibroblasts and adipocytes caveolae display a characteristic striated coat that is easily distinguished from a clathrin lattice (ROTHBERG et al. 1992). We also found that N2a cells do not express caveolin, a protein component of the caveolar coat, or its mRNA. Our results are consistent with the observation that morphologically identifiable caveolae, as well as caveolin mRNA and protein, are most abundant in endothelial cells, fibroblasts, and smooth muscle cells, and may not be present in neurons or neuronal cell lines (ROTHBERG et al. 1992; GLENNEY 1992).

5 Structural Determinants of chPrP Endocytosis

Since chPrP and other PrP^C forms lack a cytoplasmic domain, they cannot interact directly with the intracellular components of coated pits such as adaptor molecules and clathrin. We have therefore begun to explore the structural features of the chPrP molecule that mediate its endocytosis via clathrin-coated pits. Our results indicate that the polypeptide chain of chPrP, which is entirely extracellular, plays a critical role (SHYNG et al. 1995). This conclusion is based upon a biochemical and morphological analysis of deleted and chimeric forms of chPrP.

We have found that N-terminal deletions of the chPrP molecule reduces the amount of protein endocytosed in a cell-surface iodination assay (Fig. 4). The effect of these deletions is graded, with removal of 17, 41, and 67 amino acids producing a progressively larger reduction in the amount of the protein that is internalized. Each of the deleted proteins is expressed on the cell surface, where it is attached by a GPI anchor, suggesting that the anchor itself is not sufficient to signal internalization. Decay-accelerating factor (DAF), an unrelated GPI-anchored protein (LUBLIN and ATKINSON 1989), is poorly internalized in N2a cells, indicating that distinctive features of the chPrP polypeptide chain are responsible for endocytic targeting.

We also found that N-terminally deleted forms of chPrP were less concentrated in clathrin-coated pits than the wild-type protein, as determined by quantitative electron microscope immunogold labeling. Again, larger deletions had a more pronounced effect. While wild-type chPrP was 3- to 6-fold more concentrated in coated pits than over other areas of the plasma membrane, the Δ25–41 and Δ25–91 proteins were only 1.7- and 1.1-fold more concentrated, respectively.

The graded effect of the N-terminal deletions suggested that endocytic targeting of chPrP does not depend on a single discrete feature of the primary

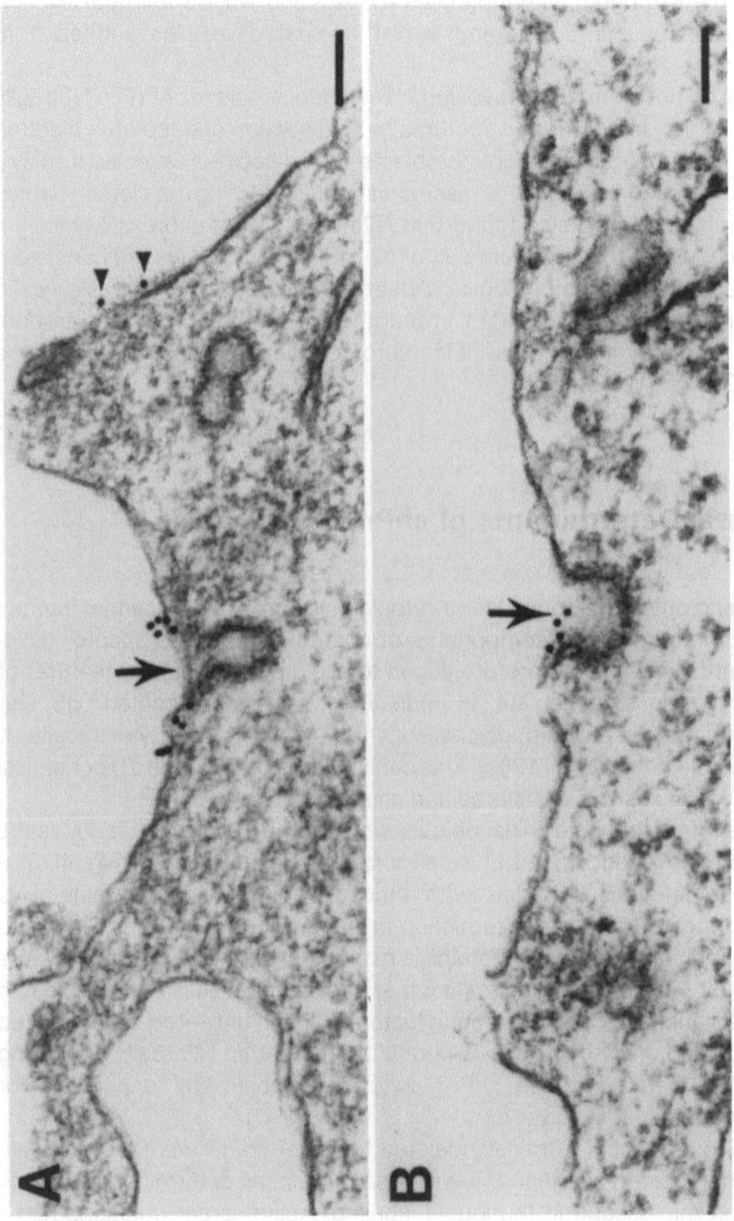
A
B

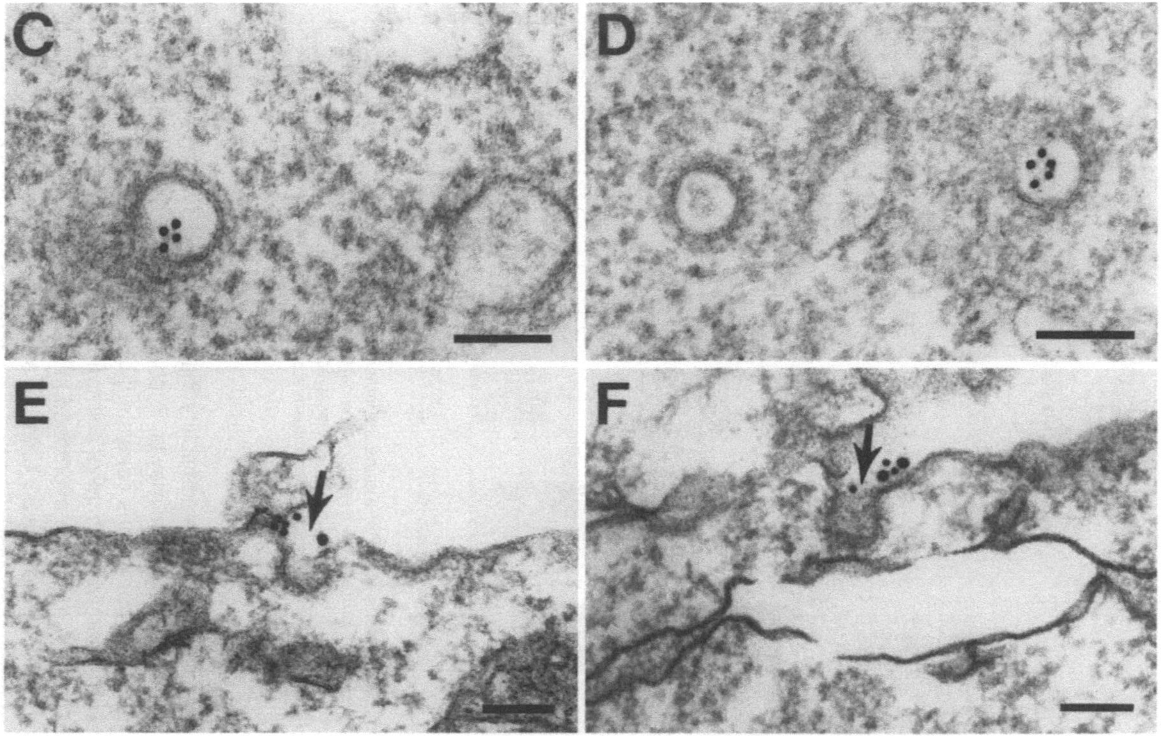

Fig. 3A–F. Immunogold localization of chPrP in N2a cells. **A, B** cells were incubated with anti-chPrP antibodies at 4°C and then fixed and processed for immunogold detection using goat anti-rabbit secondary antibodies conjugated to 10-nm gold particles. **A** Clusters of gold particles are localized near the edges of a clathrin-coated pit (*arrow*). Two isolated gold particles (*arrowheads*) are also seen over an undifferentiated area of membrane. **B** A cluster of gold particles within a coated pit (*arrow*). **C, D** Cells were incubated at 37°C for 10 min after primary antibody incubation to initiate internalization of chPrP; cells were then fixed and decorated with gold-conjugated secondary antibodies. Gold particles are seen within circular, clathrin-coated structures that represent either vesicles that have pinched off from the plasma membrane or cross-sections through deeply invaginated pits just prior to budding. **E, F** Cells were incubated at 4°C with rabbit anti-chPrP antibody and a rat monoclonal antibody against the mouse transferrin receptor. After incubation at 37°C for 10 min, the cells were fixed, and chPrP was visualized using a goat anti-rabbit antibody conjugated to 10-nm gold, and transferrin receptors with a goat anti-rat antibody conjugated to 15-nm gold. The two types of gold particles are concentrated in the same coated pits (*arrows*). *Scale bars*, 0.1µm. (From SHUNG et al. 1994, with permission)

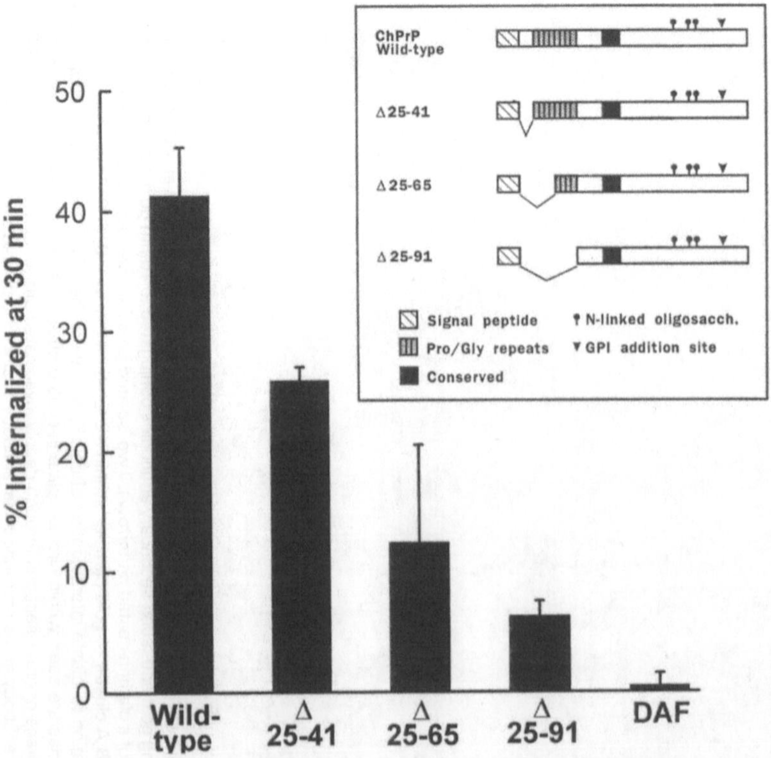

Fig. 4. Successive N-terminal deletions of chPrP reduce the efficiency of endocytosis. Stably transfected N2a cells expressing each protein were surface iodinated, and the proportion of protein internalized after 30 min was determined as in Fig. 2. Decay-accelerating factor (*DAF*) is a GPI-anchored protein unrelated to chPrP that was used as a negative control for internalization. *Inset*, the structures of the chPrP deletion mutants

sequence. Rather, our results raise the possibility that the secondary and tertiary structure of the chPrP polypeptide chain may be important in interactions with the endocytic machinery. This hypothesis is consistent with our finding that mouse PrP^C (moPrP), as well as several moPrP/chPrP chimeric proteins, are endocytosed as efficiently as wild-type chPrP when expressed in N2a cells (data not shown). Although moPrP and chPrP share only 33% amino acid sequence identity, they share the same set of predicted secondary structural domains, including a series of β-turns arising from proline- and glycine-rich peptide repeats in the N-terminal half of the molecule, and four α-helices in the C-terminal half (Bazan et al. 1987; Gasset et al. 1992). The tertiary structures of the chicken and mouse proteins may also be similar, since computer modeling predicts that they may both adopt an X-bundle conformation (Huang et al. 1994). Which of these features plays a role in endocytic targeting requires further investigation.

6 A Putative Receptor for PrP^C

The observations that chPrP is endocytosed via clathrin-coated pits, and that alterations of the extracellular polypeptide chain interfere with endocytic targeting have led us to propose the existence of a transmembrane protein which we refer to as a putative PrP^C receptor (Fig. 5). We hypothesize that the polypeptide chain of chPrP and other PrP^C forms bind to the extracellular domain of this receptor, and that a coated pit localization signal in the cytoplasmic domain of the receptor then targets the complex for endocytosis. Binding of chPrP to its receptor might be direct, as pictured in Fig. 5, or might involve intermediate proteins. In either case, deletions and truncations of the chPrP polypeptide chain would reduce its binding affinity, and so result in less efficient endocytosis. Since the alterations that we have analyzed have all been in the N-terminal half of the chPrP molecule, we would predict this region to be involved in interactions with the receptor, although a role for more C-terminal regions cannot be ruled out.

A similar mechanism seems to be responsible for endocytosis of the receptor for urokinase-type plasminogen activator, a glycolipid-anchored protein that associates with the low-density lipoprotein receptor-related protein, a large transmembrane protein that contains coated pit localization signals in its cytoplasmic domain (Nykjaer et al. 1992).

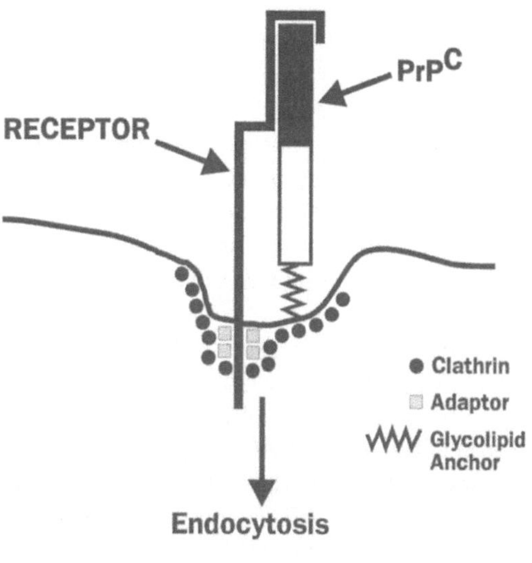

Fig. 5. Model for the function of a putative PrP^C receptor. We hypothesize that the PrP^C receptor is a transmembrane protein whose extracellular domain binds to the polypeptide chain of PrP^C, and whose cytoplasmic domain contains signals for interacting with adaptor molecules and clathrin in coated pits. The N-terminal half of PrP^C (*filled*) may interact with the receptor since deletions and truncations within this region interfere with endocytosis

7 Detergent-Resistant Complexes Contain chPrP

It has been known for a number of years that glycolipid-anchored proteins are relatively resistant to solubilization with nonionic detergents. In lysates of epithelial cells and lymphocytes, GPI-anchored proteins are found associated with large, detergent-resistant complexes that also contain sphingolipids and cholesterol, as well as cytoplasmic signaling molecules such as *src* family tyrosine kinases and heterotrimeric G proteins (Brown and Rose 1992; Sargiacomo et al. 1993). It has been postulated that these complexes are derived from caveolae since they are also enriched in the caveolar coat protein caveolin (Lisanti et al. 1993). It has also been suggested that the complexes play a role in an unusual experimental phenomenon: the ability of glycolipid-anchored proteins on the surface of lymphocytes to transmit a mitogenic signal upon cross-linking with secondary antibodies (Robinson 1991; Cinek and Horejsi 1992). The complexes have also been hypothesized to function in sorting of membrane proteins in the trans-Golgi network of polarized epithelial cells (Simons and Wadinger-Ness 1990).

Our data indicate that even though N2a cells do not display caveolae or express caveolin, Triton X-100 extracts of these cells contain detergent-resistant complexes that contain chPrP, and that display other properties of the complexes isolated from epithelial cells and lymphocytes (Gorodinsky and Harris 1995). These complexes can be isolated by floatation in sucrose gradients, presumably because of their high content of lipid and cholesterol (Fig. 6A). Because of their large size they can also be collected in the void volume during chromatography on Sepharose 4B (not shown). In addition to PrP and other glycolipid-anchored proteins such as DAF, the complexes from N2a cells contain two cytoplasmic tyrosine kinases, c-*fyn* and c-*yes*, as well as the G_α subunits G_0 and G_i (Fig. 6B). Each of these proteins is enriched 50- to 100-fold in the complexes, which contain less than 1% of the total cellular protein. The presence of a GPI anchor is critical for incorporation into the complexes, since a transmembrane version of chPrP fractionates at the bottom of the gradient with the bulk of the cellular protein.

We hypothesize that the chPrP-containing complexes which are found in detergent extracts are derived from microdomains of the plasma membrane that exist in intact cells. Consistent with this idea, the complexes are seen to consist of membranous sacs ranging in size from 200 to 800 nm when viewed by electron microscopy after thin-sectioning. These vesicles lack an obvious coat structure and are distinct in size and shape from clathrin-coated vesicles, the majority of which sediment at the bottom of the sucrose gradient, distinct from the position of the detergent-resistant complexes. Whether chPrP dissociates from these microdomains when it enters coated pits remains to be determined.

Our results raise the possibility that chPrP could be involved in signal transduction processes in neuronal cells. The detergent-resistant complexes containing chPrP are enriched in several cytoplasmic signal transducing molecules, and they exhibit tyrosine kinase activity in vitro (Fig. 6C). Interestingly, antibody-induced cross-linking of a variety of glycolipid-anchored antigens on the

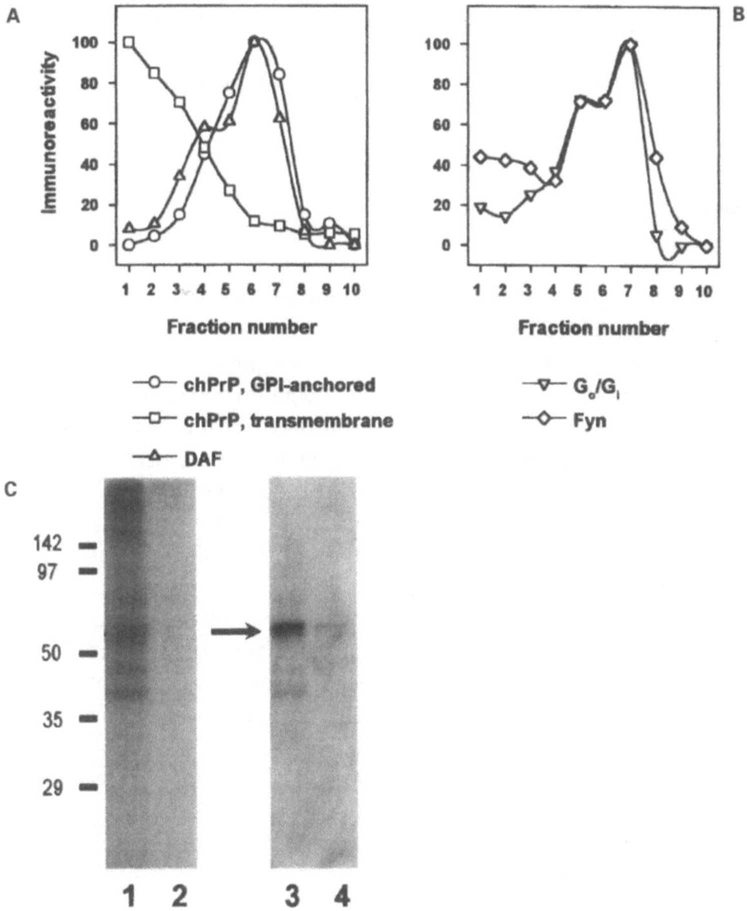

Fig. 6A–C. chPrP extracted from N2a cells is present in low-density, detergent-resistant complexes. **A** N2a cells expressing either GPI-anchored chPrP, transmembrane chPrP (chPrP residues 1–241 fused to the transmembrane and cytoplasmic domains of mannose-6-phosphate receptor), or DAF were homogenized at 4°C in a buffer containing 1% Triton X-100. The homogenate was brought to 40% sucrose and overlaid with a continuous gradient of 30%–5% sucrose. After centrifugation to equilibrium at 170 000 *g* for 16 h gradient fractions were immunoblotted for the expressed proteins. The two GPI-anchored proteins float at 15%–18% sucrose (fractions 5–7), while the transmembrane form of chPrP is found near the bottom of the gradient with the bulk of the total cellular protein. **B** ChPrP-expressing N2a cells were extracted and fractionated as described for **A**, but the gradient fractions were immunoblotted with an antibody to the tyrosine kinase c-*fyn* or with an antibody that recognizes both $G_{\alpha o}$ and $G_{\alpha i}$. These molecules float in the same region of the gradient (fractions 6,7) as chPrP (compare with **A**). A similar result was obtained for the tyrosine kinase c-*yes* (not shown). **C** Detergent-resistant complexes from chPrP-expressing N2a cells were collected at 15%–18% sucrose after equilibrium centrifugation on sucrose gradients, as described in **A** and **B**. One set of samples was incubated with $AT^{32}P$ for 0 min (*lane 2*) or 15 min (*lane 1*), and then analyzed by sodium dodecyl sulfate polyacrylamide gel electrophoresis and autoradiography. A second set of samples was incubated with nonradioactive ATP for 0 min (*lane 4*) or 15 min (*lane 3*) and then immunoblotted using an anti-phosphotyrosine antibody. Multiple proteins in the isolated complexes are phosphorylated in vitro, some of them on tyrosine. *Arrow*, a prominent group of tyrosine-phosphorylated proteins near 55–60 kDa that are likely to include *fyn*, *yes*, and perhaps other proteins of the *src* family

surface of lymphocytes stimulates tyrosine phosphorylation, increases cytoplasmic calcium levels, and initiates cell division (ROBINSON 1991). Whether PrP^C is involved in similar phenomena in neurons remains to be explored.

8 Conclusions

Taken together, our results allow us to define a detailed cellular pathway followed by chPrP, and probably other PrP^C forms (Fig. 7). Newly synthesized molecules are rapidly conjugated to their GPI anchors and are delivered to the cell surface within 30–60 min, passing through the endoplasmic reticulum and Golgi complex as are other membrane and secretory proteins. Probably in the Golgi, chPrP molecules are incorporated into membrane microdomains that are rich in sphingolipids, cholesterol, *src* family tyrosine kinases, and α-subunits of G proteins. After reaching the cell surface chPrP is endocytosed via clathrin-coated pits, a process that depends critically on distinctive features of the chPrP polypeptide chain, and that we hypothesize is mediated by a putative transmembrane PrP^C receptor. After delivery to an early endosomal compartment, a small percentage of the molecules are proteolytically cleaved within the conserved central region. Intact chPrP molecules, as well as membrane-anchored C-terminal fragments, are then returned to the cell surface, while soluble N-terminal fragments are released into the extracellular medium.

Our data are relevant to elucidating the physiological function of PrP^C, and we have suggested several possible roles for the protein, including that of a ligand, a receptor, and a signal transducing element. The identification of molecules in

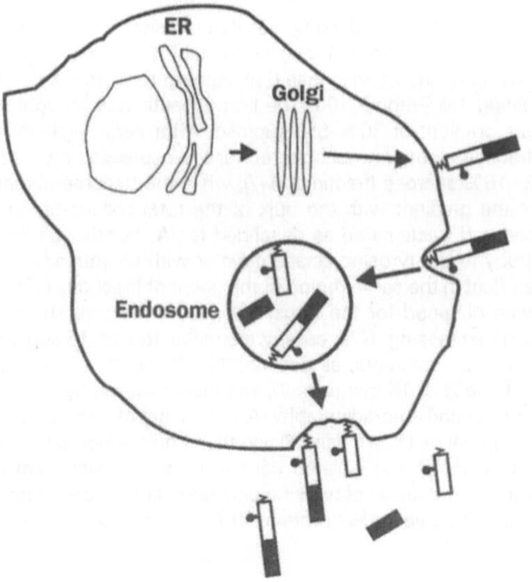

Fig. 7. Cartoon of the cellular targeting of PrP^C. See the text for details. (From SHUNG et al. 1993, with permission)

coated pits with which chPrP associates is likely to provide additional clues to the physiological function of PrP^{C}.

Our results also provide a basis for speculating about the cellular basis for generation of PrP^{Sc} during prion infection. Previous studies had suggested the involvement of an endocytic pathway in this process (CAUGHEY and RAYMOND 1991; BORCHELT et al. 1992). We now hypothesize that PrP^{C} is converted to PrP^{Sc} somewhere along the recycling route that we have defined, which involves relatively early endocytic compartments, including clathrin-coated pits and vesicles. Further, we suggest that once formed, PrP^{Sc} fails to recycle and is delivered to more distal endocytic compartments such as lysosomes. It will be interesting to determine whether PrP molecules carrying mutations linked to familial prion diseases follow a pathway different from that outlined here.

Further exploration of the mechanism of prion replication will require the application of a variety of cell biological techniques, including immunocytochemical staining and subcellular fractionation, to cultured cells expressing mutant and infectious forms of PrP. Elucidation of the cellular pathways involved may allow the design of therapeutic strategies for blocking the infectious process.

Acknowledgments. Work in the laboratory is supported by grants to D.A.H. from the NIH (NS30137), March of Dimes, and Alzheimer's Association (Richard F. Bristor Investigator-Initiated Research Grant). S.L. is supported by a Postdoctoral Research Fellowship for Physicians from the Howard Hughes Medical Institute, and S.-L.S. by a postdoctoral training grant from the NIH (T32 NS07071).

References

Anderson RGW (1993) Plasmalemmal caveolae and GPI-anchored membrane proteins. Curr Opin Cell Biol 5: 647–652

Bazan JF, Fletterick RJ, McKinley MP, Prusiner SB (1987) Predicted secondary structure and membrane topology of the scrapie prion protein. Protein Eng 1: 125–135

Borchelt DR, Taraboulos A, Prusiner SB (1992) Evidence for synthesis of scrapie prion proteins in the endocytic pathway. J Biol Chem 267: 16188–16199

Bretscher MS, Thomson JN, Pearse BMF (1980) Coated pits act as molecular filters. Proc Natl Acad Sci USA 77: 4156–4159

Brown DA, Rose JK (1992) Sorting of GPI-anchored proteins to glycolipid-enriched membrane subdomains during transport to the apical cell surface. Cell 68: 533–544

Caughey B, Raymond GJ (1991) The scrapie-associated form of PrP is made from a cell surface precursor that is both protease- and phospholipase-sensitive. J Biol Chem 266: 18217–18223

Caughey BW, Dong A, Bhat KS, Ernst D, Hayes SF, Caughey WS (1991) Secondary structure analysis of the scrapie-associated protein PrP 27–30 in water by infared spectroscopy. Biochemistry 30: 7672–7680

Cinek T, Horejsi V (1992) The nature of large noncovalent complexes containing glycosyl-phosphatidylinositol-anchored membrane glycoproteins and protein tyrosine kinases. J Immunol 149: 2262–2270

Collinge J, Whittington MA, Sidle KCL, Smith CJ, Palmer MS, Clarke AR, Jefferys JGR (1994) Prion protein is necessary for normal synaptic transmission. Nature 370: 295–297

Gabriel J-M, Oesch B, Kretzschmar H, Scott M, Prusiner SB (1992) Molecular cloning of a candidate chicken prion protein. Proc Natl Acad Sci USA 89: 9097–9101

Gasset M, Baldwin MA, Lloyd DH, Gabriel J-M, Holtzman DM, Cohen F, Fletterick R, Prusiner SB (1992) Predicted α-helical regions of the prion protein when synthesized as peptides form amyloid. Proc Natl Acad Sci USA 89: 10940–10944

Glenney JR (1992) The sequence of human caveolin reveals identity with VIP21, a component of transport vesicles. FEBS Lett 314: 45–48

Gorodinsky A, Harris DA (1995) Glycolipid-anchored proteins in neuroblastoma cells form detergent-resistant complexes without caveolin. J Cell Biol 129: 619–627

Harris DA, Lele P, Snider WD (1993a) Localization of the mRNA for a chicken prion protein by in situ hybridization. Proc Natl Acad Sci USA 90: 4309–4313

Harris DA, Huber M, van Dijken P, Shyng S-L, Chait B, Wang R (1993b) Processing of a cellular prion protein: identification of N- and C-terminal cleavage sites. Biochemistry 32: 1009–1016

Heuser JE, Anderson RGW (1989) Hypertonic media inhibit receptor-mediated endocytosis by blocking clathrin-coated pit formation. J Cell Biol 108: 389–400

Huang Z, Gabriel J-M, Baldwin MA, Fletterick RJ, Prusiner SB, Cohen FE (1994) Proposed three-dimensional structure for the cellular prion protein. Proc Natl Acad Sci USA 91: 7139–7143

Keller G-A, Siegel MW, Caras IW (1992) Endocytosis of glycophospholipid-anchored and transmembrane forms of CD4 by different endocytic pathways. EMBO J 11: 863–874

Lehmann S, Shyng S-L, Harris DA (1994) Cellular processing of normal and mutant mammalian prion proteins. Neurobiol Aging 15 [Suppl 1]: S87–88

Lisanti MP, Tang Z, Sargiacomo M (1993) Caveolin forms a hetero-oligomeric protein complex that interacts with an apical GPI-linked protein: implications for the biogenesis of caveolae. J Cell Biol 123: 595–604

Lublin DM, Atkinson JP (1989) Decay-accelerating factor: biochemistry, molecular biology, and function. Annu Rev Immunol 7: 35–58

Manson J, West JD, Thomson V, McBride P, Kaufman MH, Hope J (1992) The prion protein gene: a role in mouse embryogenesis? Development 115: 117–122

Massagué J (1990) Transforming growth factor-α: a model for membrane-anchored growth factors. J Biol Chem 265: 21393–21396

McKinley MP, Taraboulos A, Kenaga L, Serban D, Stieber A, DeArmond SJ, Prusiner SB, Gonatas N (1991) Ultrastructural localization of scrapie prion proteins in cytoplasmic vesicles of infected cultured cells. Lab Invest 65: 622–630

Nykjaer A, Petersen CM, Moller B, Jensen PH, Moestrup SK, Holtet TL, Etzerodt M, Thogersen HC, Munch M, Andreasen PA, Gliemann J (1992) Purified α_2-macroglobulin receptor/LDL receptor related protein binds urokinase plasminogen activator inhibitor type-1 complex: evidence that the α_2-macroglobulin receptor mediates cellular degradation of urokinase receptor-bound complexes. J Biol Chem 267: 14543–14546

Oesch B, Westaway D, Walchli M, McKinley MP, Kent SBH, Aebersold R, Barry RA, Tempst P, Teplow DB, Hood L, Prusiner SB, Weissmann C (1985) A cellular gene encodes scrapie PrP 27–30 protein. Cell 40: 735–746

Pan K-M, Baldwin M, Nguyen J, Gasset M, Serban A, Groth D, Mehlhorn I, Huang Z, Fletterick RJ, Cohen FE, Prusiner SB (1993) Conversion of α-helices into β-sheets features in the formation of the scrapie prion proteins. Proc Natl Acad Sci USA 90: 10962–10966

Prusiner SB, DeArmond SJ (1994) Prion diseases and neurodegeneration. Annu Rev Neurosci 17: 311–339

Prusiner SB, Scott M, Foster D, Pan K-M, Groth D, Mirenda C, Torchia M, Yang S-L, Serban D, Carlson GA, Hoppe PC, Westaway D, DeArmond SJ (1990) Transgenetic studies implicate interactions between homologous PrP isoforms in scrapie prion replication. Cell 63: 673–686

Robinson PJ (1991) Phosphatidylinositol membrane anchors and T-cell activation. Immunol Today 37: 35–41

Rogers M, Yehiely F, Scott M, Prusiner SB (1993) Conversion of truncated and elongated prion proteins into the scrapie isoform in cultured cells. Proc Natl Acad Sci USA 90: 3182–3186

Rothberg KG, Ying Y-S, Kolhouse JF, Kamen BA, Anderson RGW (1990) The glycophospholipid-linked folate receptor internalizes folate without entering the clathrin-coated pit endocytic pathway J Cell Biol 110: 637–649

Rothberg KG, Heuser J, Donzell WC, Ying Y-S, Glenny JR, Anderson RGW (1992) Caveolin, a protein component of caveolae membrane coats. Cell 68: 673–682

Safar J, Roller PP, Gajdusek DC, Gibbs CJ (1993) Conformational transitions, dissociation, and unfolding of scrapie amyloid (prion) protein. J Biol Chem 268: 20276–20284

Sargiacomo M, Sudol M, Tang Z-L, Lisanti MP (1993) Signal transducing molecules and glycosyl-phosphatidylinositol-linked proteins form a caveolin-rich insoluble complex in MDCK cells. J Cell Biol 122: 789–807

Shyng S-L, Huber MT, Harris DA (1993) A prion protein cycles between the cell surface and an endocytic compartment in cultured neuroblastoma cells. J Biol Chem 268: 15922–15928

Shyng S-L, Heuser JE, Harris DA (1994) A glycolipid-anchored prion protein is endocytosed via clathrin-coated pits. J Cell Biol 125: 1239–1250
Shyng S-L, Moulder KL, Lesko A, Harris DA (1995) The N-terminal of a glycolipid-anchored prion protein is essential for its endocytosis via clathrin-coated pits. J Biol Chem 270: 14793–14800
Simons K, Wandinger-Ness A (1990) Polarized sorting in epithelia. Cell 62: 207–210
Taraboulos A, Serban D, Prusiner SB (1990a) Scrapie prion proteins accumulate in the cytoplasm of persistently infected cultured cells. J Cell Biol 110: 2117–2132
Taraboulos A, Rogers M, Borchelt DR, McKinley MP, Scott M, Serban D, Prusiner SB (1990b) Acquisition of protease resistance by prion proteins in scrapie-infected cells does not require asparagine-linked glycosylation. Proc Natl Acad Sci USA 87: 8262–8266
Trowbridge IS, Collawn JF, Hopkins CR (1993) Signal-dependent membrane protein trafficking in the endocytic pathway. Annu Rev Cell Biol 9: 129–161

Transgenetics and Gene Targeting in Studies of Prion Diseases*

M.R.D. Scott[1], G.C. Telling[1], and S.B. Prusiner[1,2]

1 Introduction

Prions cause a group of human and animal neurodegenerative diseases which are now classified together because their etiology and pathogenesis involve modification of the prion protein (PrP) (Prusiner 1991). Prion diseases are manifest as infectious, genetic, and sporadic disorders (Table 1). These diseases can be transmitted among mammals by the infectious particle designated "prion" (Prusiner 1982). Despite intensive searches over the past three decades no nucleic acid has been found within prions (Alper et al. 1966, 1967; Hunter 1972; Riesner et al. 1992); however, a modified isoform of the host-encoded PrP

---

*This chapter also appears in Vol.206: Transgenic Models of Human Viral and Immunological Disease (F.V. Chisari and M.B.A. Oldstone)

[1]Department of Neurology, University of California, San Francisco, CA 94143-0518, USA
[2]Department of Biochemistry and Biophysics, University of California, San Francisco, CA 94143-0518, USA

Table 1. Human prion diseases

Disease	Etiology
Kuru	Infection
Creutzfeldt-Jakob disease	
Iatrogenic	Infection
Sporadic	Unknown
Familial	PrP mutation
Gerstmann-Sträussler-Scheinker disease	PrP mutation
Fatal familial insomnia	PrP mutation

designated PrP^{Sc} is essential for infectivity (Büeler et al. 1993; Prusiner 1991; Prusiner et al. 1983, 1993b,c). In fact, considerable experimental data argue that prions are composed exclusively of PrP^{Sc}. Earlier terms used to describe the prion diseases include: transmissible encephalopathies, spongiform encephalopathies, and slow virus diseases (Gajdusek 1977, 1985; Sigurdsson 1954).

The quartet of human (Hu) prion diseases are frequently referred to as kuru, Creutzfeldt-Jakob disease (CJD), Gerstmann-Sträussler-Scheinker (GSS) disease, and fatal familial insomnia (FFI). Kuru was the first of the human prion diseases to be transmitted to experimental animals, and it has often been suggested that kuru spread among the Fore people of Papua New Guinea by ritualistic cannibalism (Gajdusek 1977; Gajdusek et al. 1966). The experimental and presumed human to human transmission of kuru led to the belief that prion diseases are infectious disorders caused by unusual viruses similar to those causing scrapie in sheep and goats. Yet, the occurrence of CJD in families, first reported almost 70 years ago (Kirschbaum 1924; Meggendorfer 1930), was perplexing to say the least. The significance of familial CJD remained unappreciated until mutations in the protein coding region of the PrP gene on the short arm of chromosome 20 were discovered (Hsiao et al. 1989; Prusiner 1994; Sparkes et al. 1986). The earlier finding that brain extracts from patients who had died of familial prion diseases inoculated into experimental animals often transmit disease, posed a conundrum that was resolved with the genetic linkage of these diseases to mutations of the PrP gene (Masters et al. 1981; Prusiner 1989; Tateishi et al. 1992).

The most common form of prion disease is sporadic CJD. Many attempts to show that the sporadic prion diseases are caused by infection have been unsuccessful (Brown et al. 1987; Cousens et al. 1990; Harries-Jones et al. 1988; Malmgren et al. 1979). The discovery that inherited prion diseases are caused by germline mutation of the PrP gene raised the possibility that sporadic forms of these diseases might result from a somatic mutation (Prusiner 1989). The discovery that PrP^{Sc} is formed from the cellular isoform of the prion protein (PrP^{C}) by a posttranslational process (Borchelt et al. 1990), and that overexpression of wild-type (wt) PrP transgenes produces spongiform degeneration and infectivity de novo (Westaway et al. 1994b) has raised the possibility that sporadic prion diseases result from the spontaneous conversion of PrP^{C} into PrP^{Sc}.

CJD has a worldwide incidence of approx. one case per 10^6 population (Masters et al. 1978). Fewer than 1% of CJD cases are infectious, and all of those appear to be iatrogenic. Between 10% and 15% of prion disease cases are inherited while the remaining cases are sporadic. Kuru was once the most common cause of death among New Guinea women in the Fore region of the Highlands (GAJDUSEK et al. 1966; GAJDUSEK and ZIGAS 1957; GAJDUSEK and ZIGAS 1959) but has virtually disappeared with the cessation of ritualistic cannibalism (ALPERS 1987). Patients with CJD frequently present with dementia, but approx. 10% of patients exhibit cerebellar dysfunction initially. Patients with either kuru or GSS usually present with ataxia while those with FFI manifest insomnia and autonomic dysfunction (BROWN 1992; HSIAO and PRUSINER 1990; MEDORI et al. 1992b).

PrP^{CJD} has been found in the brains of most patients who died of prion disease. The term PrP^{CJD} is preferred by some investigators when referring to the abnormal isoform of HuPrP in human brain. Here, PrP^{Sc} is used interchangeably with PrP^{CJD}. PrP^{Sc} is always used after human CJD prions have been passaged into an experimental animal since the nascent PrP^{Sc} molecules are produced from host PrP^{C}, and the PrP^{CJD} in the inoculum serves only to initiate the process. In the brains of some patients with inherited prion diseases as well as transgenic (Tg) mice expressing mouse (Mo) PrP with the human GSS point mutation (Pro→Leu), detection of PrP^{Sc} has been problematic despite clinical and neuropathologic hallmarks of neurodegeneration (HSIAO et al. 1990, 1994). Presumably, neurodegeneration is due at least in part to the abnormal metabolism of mutant PrP (HSIAO et al. 1990). Of note, horizontal transmission of neurodegeneration from the brains of patients with inherited prion diseases to inoculated rodents has been less frequent than with sporadic cases (TATEISHI et al. 1992). Whether this distinction between transmissible and nontransmissible inherited prion diseases will persist is unclear. Tg mice expressing a chimeric Hu/Mo PrP gene have been found to be highly susceptible to Hu prions from sporadic and iatrogenic CJD cases (TELLING et al. 1994). These Tg(MHu2M) mice should make the use of apes and monkeys for the study of human prion diseases unnecessary and allow for tailoring the PrP^{C} translated from the transgene to match the sequence of the PrP^{CJD} in the inoculum. Other Tg mouse studies have demonstrated that PrP^{Sc} in the inoculum interacts preferentially with homotypic PrP^{C} during the propagation of prions (PRUSINER et al. 1990; SCOTT et al. 1993). PrP^{C} is the cellular isoform of the prion protein which has been identified in all mammals and birds examined to date; PrP^{C} is anchored to the external surface of cells by a glycolipid moiety and its function is unknown (STAHL et al. 1987).

Scrapie is the most common natural prion disease of animals. An investigation into the etiology of scrapie followed the vaccination of sheep for looping ill virus with formalin-treated extracts of ovine lymphoid tissue unknowingly contaminated with scrapie prions (GORDON 1946). Two years later more than 1500 sheep developed scrapie from this vaccine. While the transmissibility of experimental scrapie became well established, the spread of natural scrapie within and among flocks of sheep remained puzzling. Parry argued that host genes were responsible for the development of scrapie in sheep. He was convinced that natural scrapie is a genetic disease which could be eradicated by proper breeding protocols (PARRY 1962, 1983). He considered its transmission by inoculation of importance primarily for laboratory studies and communicable infection of little consequence in nature. Other investigators view natural scrapie as an infectious disease and argue that host genetics only modulates susceptibility to an endemic infectious agent (DICKINSON et al. 1965).

The offal of scrapied sheep in the United Kingdom is thought to be responsible for the current epidemic of bovine spongiform encephalopathy (BSE) or mad cow disease (WILESMITH et al. 1992b). Prions in the offal from scrapie-infected sheep appear to have survived rendering which produced meat and bone meal (MBM). After BSE was recognized, MBM produced from domestic animals was banned from further use. Since 1986 when BSE was first recognized, more than 130 000 cattle have died of BSE. Whether humans develop CJD after consuming beef from cattle with BSE prions is of considerable concern.

The fundamental event in prion diseases seems to be a conformational change in PrP. All attempts to identify a posttranslational chemical modification that distinguishes PrP^{Sc} from PrP^{C} have been unsuccessful to date (STAHL et al. 1993). PrP^{C} contains approx. 45% α-helix and is virtually devoid of β-sheet (PAN et al. 1993). Conversion to PrP^{Sc} creates a protein which contains approx. 30% α-helix and 45% β-sheet. The mechanism by which PrP^{C} is converted into PrP^{Sc} remains unknown, but PrP^{C} appears to bind to PrP^{Sc} to form an intermediate complex during the formation of nascent PrP^{Sc}.

Prions differ from all other known infectious pathogens in several respects. First, prions do not contain a nucleic acid genome which codes for their progeny. Viruses, viroids, bacteria, fungi, and parasites all have nucleic acid genomes that code for their progeny. Second, the only known component of the prion is a modified protein that is encoded by a cellular gene. Third, the major, and possibly only, component of the prion is PrP^{Sc} which is a pathogenic conformer of PrP^{C}.

As our knowledge of the prion diseases increases, and more is learned about the molecular and genetic characteristics of prion proteins, these disorders will undoubtedly undergo modification with respect to their classification. Indeed, the discovery of the PrP and the identification of pathogenic PrP gene mutations have already forced us to view these illnesses from perspectives not previously imagined.

2 Transgenetics and Gene Targeting

While transgenetic studies have yielded a wealth of new knowledge about infectious, genetic, and sporadic prion diseases, the laborious production of Tg mice limits the number of studies that can be performed. The relatively long gestation period of mice coupled with the need to perform microinjections of fertilized embryos prevents the creation of the very large numbers of different Tg mice that would yield the greatest amount of new information. Assays that permit screening of a multitude of possible phenotypes in genetic experiments are generally the most informative. While the limited number of mice expressing different transgenes is definitely a liability, experiments with Tg mice expressing foreign and mutant PrP molecules have been extraordinarily useful in advancing our understanding of prion biology. It is important to stress that transgenetic studies can readily yield an incomplete, and sometimes erroneous interpretation of the data if the number of lines of mice examined expressing a particular construct is inadequate. Defining an adequate number of lines is difficult, but certainly comparisons of lines expressing high and low levels of a given PrP transgene have proven to be quite helpful (Hsiao et al. 1994; Prusiner et al. 1990).

2.1 Species Barriers for Transmission of Prion Diseases

The passage of prions between species is a stochastic process characterized by prolonged incubation times (Pattison 1965, 1966; Pattison and Jones 1967). Prions synthesized de novo reflect the sequence of the host PrP gene and not that of the PrP^{Sc} molecules in the inoculum (Bockman et al. 1987).

On subsequent passage in a homologous host, the incubation time shortens to that recorded for all subsequent passages, and it becomes a nonstochastic process. The species barrier concept is of practical importance in assessing the risk for humans of developing CJD after consumption of scrapie-infected lamb or BSE-infected beef (Goldmann et al. 1991; Hope et al. 1988; Prusiner et al. 1993a; Wilesmith et al. 1992a,b).

To test the hypothesis that differences in PrP gene sequences are responsible for the species barrier, Tg mice expressing SHaPrP were constructed (Prusiner et al. 1990; Scott et al. 1989). The PrP genes of Syrian hamsters and mice encode proteins differing at 16 positions. Incubation times in four lines of Tg(SHaPrP) mice inoculated with Mo prions were prolonged compared to those observed for non-Tg, control mice (Fig. 1A). Inoculation of Tg(SHaPrP) mice with SHa prions demonstrated abrogation of the species barrier resulting in abbreviated incubation times due to a nonstochastic process (Fig. 1B; Prusiner et al. 1990; Scott et al. 1989). The length of the incubation time after inoculation with SHa prions was inversely proportional to the level of $SHaPrP^{C}$ in the brains of Tg(SHaPrP) mice (Fig. 1B, 1C) (Prusiner et al. 1990). $SHaPrP^{Sc}$ levels in the brain of clinically ill mice were similar in all four Tg(SHaPrP) lines inoculated with SHa prions (Fig. 1D). Bioassays of brain extracts from clinically ill Tg(SHaPrP) mice inoculated with Mo prions revealed that only Mo prions, but no SHa prions were produced (Fig. 1E). Conversely, inoculation of Tg(SHaPrP) mice with SHa prions led to the synthesis only of SHa prions (Fig. 1F). Thus, the de novo synthesis of prions is species specific and reflects the genetic origin of the inoculated prions. Similarly, the neuropathology of Tg(SHaPrP) mice is determined by the genetic origin of prion inoculum. Mo prions injected into Tg(SHaPrP) mice produced a neuropathology characteristic of mice with scrapie. A moderate degree of vacuolation in both the gray and white matter was found while amyloid plaques were rarely detected (Fig. 1G; Table 2). Inoculation of Tg(SHaPrP) mice with SHa prions produced

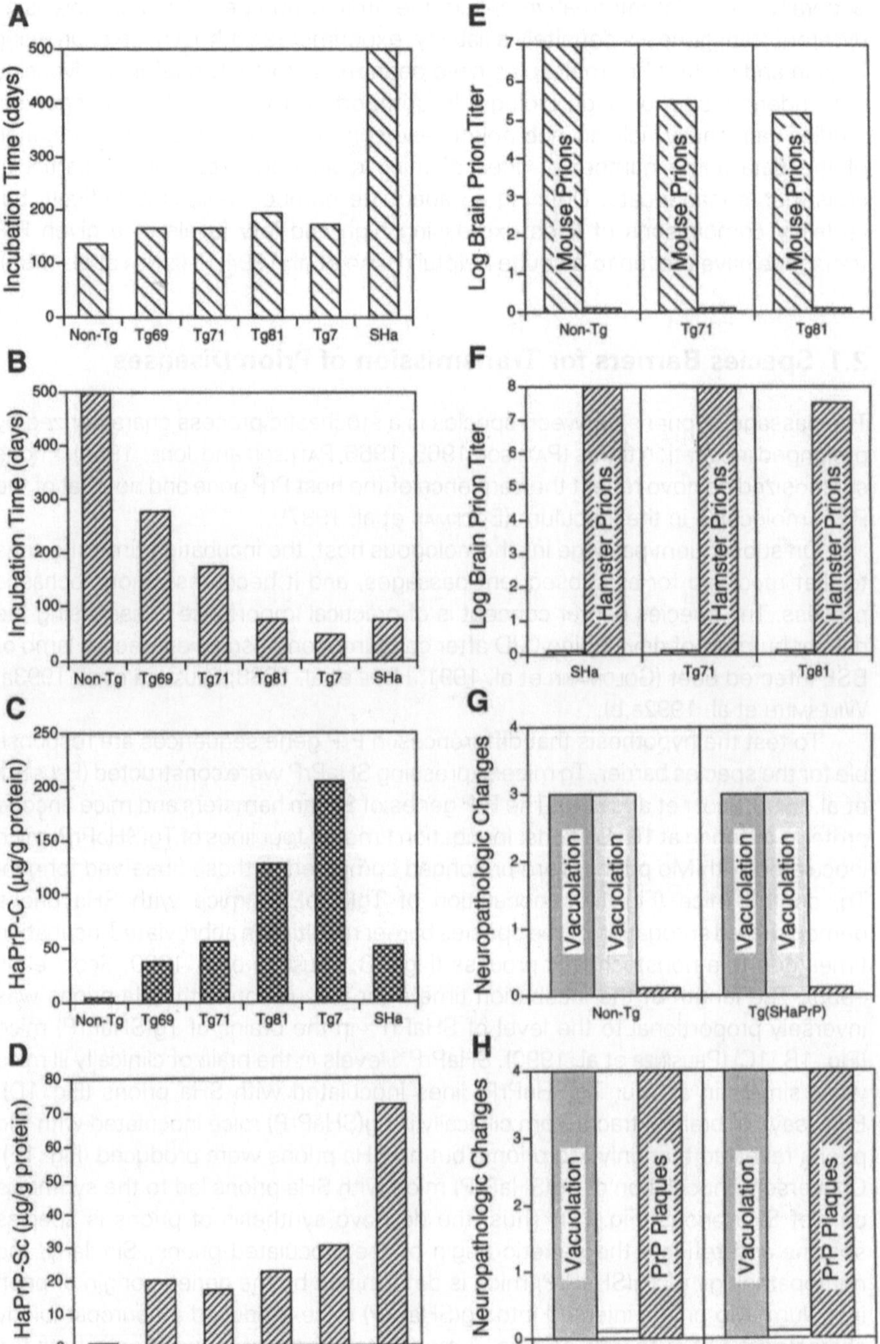

A
Incubation Time (days)
0
100
200
300
400
500
Non-Tg
Tg69
Tg71
Tg81
Tg7
SHa
B
Incubation Time (days)
0
100
200
300
400
500
Non-Tg
Tg69
Tg71
Tg81
Tg7
SHa
C
HaPrP-C (µg/g protein)
0
50
100
150
200
250
Non-Tg
Tg69
Tg71
Tg81
Tg7
SHa
D
HaPrP-Sc (µg/g protein)
0
10
20
30
40
50
60
70
80
Non-Tg
Tg69
Tg71
Tg81
Tg7
SHa
E
Log Brain Prion Titer
0
1
2
3
4
5
6
7
Mouse Prions
Mouse Prions
Mouse Prions
Non-Tg
Tg71
Tg81
F
Log Brain Prion Titer
0
1
2
3
4
5
6
7
8
Hamster Prions
Hamster Prions
Hamster Prions
SHa
Tg71
Tg81
G
Neuropathologic Changes
0
1
2
3
4
Vacuolation
Vacuolation
Vacuolation
Vacuolation
Non-Tg
Tg(SHaPrP)
H
Neuropathologic Changes
0
1
2
3
4
Vacuolation
PrP Plaques
Vacuolation
PrP Plaques
SHa
Tg(SHaPrP)

Table 2. Species-specific prion inocula determine the distribution of spongiform change and deposition of PrP amyloid plaques in transgenic mice (from PRUSINER 1992)

Animal	SHa prions					Mo prions			
		Spongiform change[a]		PrP plaques[b]			Spongiform change[a]		PrP plaques[b]
	n	Gray	Wht	Frequency	Diameter[c]	*n*	Gray	Wht	Frequency
Non-Tg		ND		ND		10	+	+	–
Tg 69	6	+	–	Numerous	6.5 ± 3.1 (389)	2	+	+	–
Tg 71	5	+	–	Numerous	8.1 ± 3.6 (345)	2	+	+	–
Tg 81	7	+	–	Numerous	8.3 ± 3.0 (439)	3	+	+	Few
Tg 7	3	+[d]	–	Numerous	14.0 ± 8.3 (19)	4	+	+	–
SHa	3	+	–	Numerous	5.7 ± 2.7 (247)		ND		ND

n, Number of brains examined; ND, not determined; +, present; –, not found.
[a] Spongiform change evaluated in hippocampus, thalamus, cerebral cortex, and brainstem for gray matter and the deep cerebellum for white matter.
[b] Plaques in the subcallosal region were stained with SHaPrP mAb 13A5, anti-PrP rabbit antisera R073, and trichrome stain.
[c] Mean diameter of PrP plaques given in microns ± standard error with the number of observations in parentheses.
[d] Focal: confirmed to the dorsal nucleus of the raphe.

intense vacuolation of the gray matter, sparing of the white matter, and numerous SHaPrP amyloid plaques characteristic of Syrian hamsters with scrapie (Fig. 1H).

2.2 Overexpression of wtPrP Transgenes

During transgenetic studies we discovered that uninoculated older mice harboring high copy numbers of wtPrP transgenes derived from Syrian hamsters, sheep, and PrP-B mice spontaneously developed truncal ataxia, hind-limb paralysis, and

Fig. 1A–H. Transgenic mice expressing Syrian hamster (*SHa*) prion protein exhibit species-specific scrapie incubation times, infectious prion synthesis and neuropathology. **A** Scrapie incubation times in nontransgenic mice and four lines of Tg mice expressing SHaPrP and Syrian hamsters inoculated intracerebrally with approx. 10^6 ID_{50} U Chandler Mo prions serially passaged in Swiss mice. The four lines of Tg mice have different numbers of transgene copies: Tg69 and 71 mice have 2-Y copies of the SHaPrP transgene, whereas Tg81 have 30–50, and Tg7 mice have more than 60. Incubation times are number of days from inoculation to onset of neurologic dysfunction. **B** Scrapie incubation times in mice and hamsters inoculated with approx. 10^7 ID_{50} U Sc237 prions serially passaged in Syrian hamsters and as described in **A**. **C** Brain $SHaPrP^{C}$ in Tg mice and hamsters. $SHaPrP^{C}$ levels were quantitated by an enzyme-linked immunoassay. **D** Brain $SHaPrP^{Sc}$ in Tg mice and hamsters. Animals were killed after exhibiting clinical signs of scrapie. $SHaPrP^{Sc}$ levels were determined by immunoassay. **E** Prion titers in brains of clinically ill animals after inoculation with Mo prions. Brain extracts from Non-Tg, Tg71, and Tg81 mice were bioassayed for prions in mice (*left*) and hamsters (*right*). **F** Prion titers in brains of clinically ill animals after inoculation with SHa prions. Brain extracts from Syrian hamsters as well as Tg71 and Tg81 mice were bioassayed for prions in mice (*left*) and hamsters (*right*). **G** Neuropathology in Non-Tg mice and Tg(SHaPrP) mice with clinical signs of scrapie after inoculation with Mo prions. Vacuolation in gray (*left*) and white matter (*center*); PrP amyloid plaques (*right*). Vacuolation score: *0*, none; *1*, rare; *2*, modest; *3*, moderate; *4*, intense. **H** Neuropathology in Syrian hamsters and Tg mice inoculated with SHa prions. Degree of vacuolation and frequency of PrP amyloid plaques as described in **G**. (Adapted from PRUSINER 1991)

tremors (WESTAWAY et al. 1994a). These Tg mice exhibited a profound necrotizing myopathy involving skeletal muscle, a demyelinating polyneuropathy, and focal vacuolation of the CNS. Development of disease was dependent on transgene dosage. For example, Tg(SHaPrP$^{+/+}$)7 mice homozygous for the SHaPrP transgene array regularly developed disease between 400 and 600 days of age, while hemizygous Tg(SHaPrP$^{+/0}$)7 mice also developed disease, but after more than 650 days.

Attempts to demonstrate PrP^{Sc} in either muscle or brain were unsuccessful but transmission of disease with brain extracts from Tg(SHaPrP$^{+/+}$)7 mice inoculated into Syrian hamsters did occur. These Syrian hamsters had PrP^{Sc} as detected by immunoblotting and spongiform degeneration (D. GROTH and S.B PRUSINER, unpublished data). Serial passage with brain extracts from these animals to recipients was observed. De novo synthesis of prions in Tg(SHaPrP$^{+/+}$)7 mice overexpressing wtSHaPrP^{C} provides support for the hypothesis that sporadic CJD does not result from infection but rather is a consequence of the spontaneous, although rare, conversion of PrP^{C} into PrP^{Sc}. Alternatively, a somatic mutation in which mutant SHaPrP^{C} is spontaneously converted into PrP^{Sc} as in the inherited prion diseases could also explain sporadic CJD. These findings as well as those described below for Tg(MoPrP-P101L) mice argue that prions are devoid of foreign nucleic acid, in accord with many earlier studies that use other experimental approaches as described above.

2.3 Ablation of the PrP Gene

Ablation of the PrP gene in Tg (Prn-p$^{0/0}$) mice has, unexpectedly, not affected the development of these animals (BÜELER et al. 1992). In fact, they are healthy at almost 2 years of age. Prn-p$^{0/0}$ mice are resistant to prions (Fig. 2) and do not propagate scrapie infectivity (BÜELER et al. 1993; PRUSINER et al. 1993b).

Prn-p$^{0/0}$ mice crossed with Tg(SHaPrP) mice were rendered susceptible to SHa prions but remained resistant to Mo prions (BÜELER et al. 1993; PRUSINER et al. 1993b). Since the absence of PrP^{C} expression does not provoke disease, it is likely that scrapie and other prion diseases are a consequence of PrP^{Sc} accumulation rather than an inhibition of PrP^{C} function (BÜELER et al. 1992).

Mice heterozygous (Prn-p$^{0/+}$) for ablation of the PrP gene had prolonged incubation times when inoculated with Mo prions (Fig. 2) (PRUSINER et al. 1993b). The Prn-p$^{0/+}$ mice developed signs of neurologic dysfunction 400–460 days after inoculation. These findings are in accord with studies on Tg(SHaPrP) mice in which increased SHaPrP expression was accompanied by diminished incubation times (Fig. 1B) (PRUSINER et al. 1990).

Since Prn-p$^{0/0}$ mice do not express PrP^{C}, we reasoned that they might more readily produce α-PrP antibodies. Prn-p$^{0/0}$ mice immunized with Mo or SHa prion rods produced α-PrP antisera which bound Mo, SHa, and Hu PrP (PRUSINER et al. 1993b). These findings contrast with earlier studies in which α-MoPrP antibodies could not be produced in mice presumably because the mice had been rendered

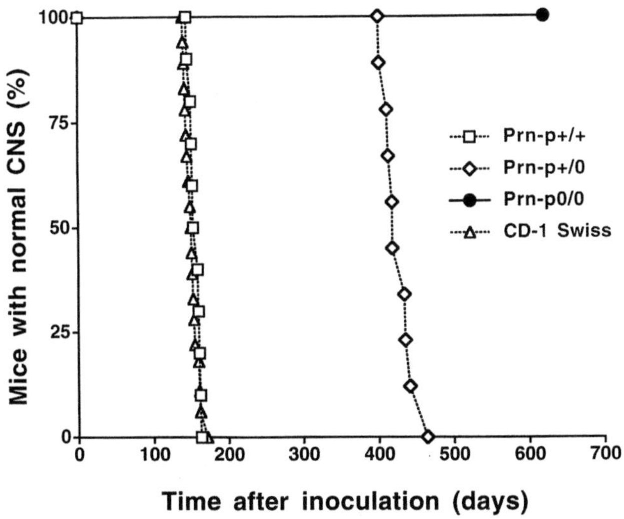

Fig. 2. Incubation times in PrP gene ablated Prn-$p^{0/+}$ and Prn-$p^{0/0}$ mice as well as wt Prn-$p^{+/+}$ and CD-1 mice inoculated with RML mouse prions. The RML prions were heated and irradiated at 254 nm prior to intracerebral inoculation into CD-1 Swiss mice (*open triangles*), Prn-$p^{+/+}$ mice (*open squares*), Prn-$p^{0/+}$ mice (*open diamonds*) or Prn-$p^{0/0}$ mice (*filled circle*)

tolerant by the presence of $MoPrP^C$ (Barry and Prusiner 1986; Kascsak et al. 1987; Rogers et al. 1991). That Prn-$p^{0/0}$ mice readily produce α-PrP antibodies is consistent with the hypothesis that the lack of an immune response in prion diseases is due to the fact that PrP^C and PrP^{Sc} share many epitopes. Whether Prn-$p^{0/0}$ mice produce α-PrP antibodies that specifically recognize conformational dependent epitopes present on PrP^{Sc} but absent from PrP^C remains to be determined.

2.4 Modeling of GSS in Tg(MoPrP-P101L) Mice

The codon 102 point mutation found in GSS patients was introduced into the MoPrP gene and Tg(MoPrP-P101L)H mice were created expressing high (H) levels of the mutant transgene product. The two lines of Tg(MoPrP-P101L)H mice designated 174 and 87 spontaneously developed CNS degeneration, characterized by clinical signs indistinguishable from experimental murine scrapie and neuropathology consisting of widespread spongiform morphology and astrocytic gliosis (Hsiao et al. 1990) and PrP amyloid plaques (Fig. 3) (Hsiao et al. 1994). By inference, these results contend that PrP gene mutations cause GSS, familial CJD, and FFI.

Brain extracts prepared from spontaneously ill Tg(MoPrP-P101L)H mice transmitted CNS degeneration in Tg196 mice expressing low levels of the mutant transgene product and some Syrian hamsters (Hsiao et al. 1994). Many Tg196 mice and some Syrian hamsters developed CNS degeneration between 200 and

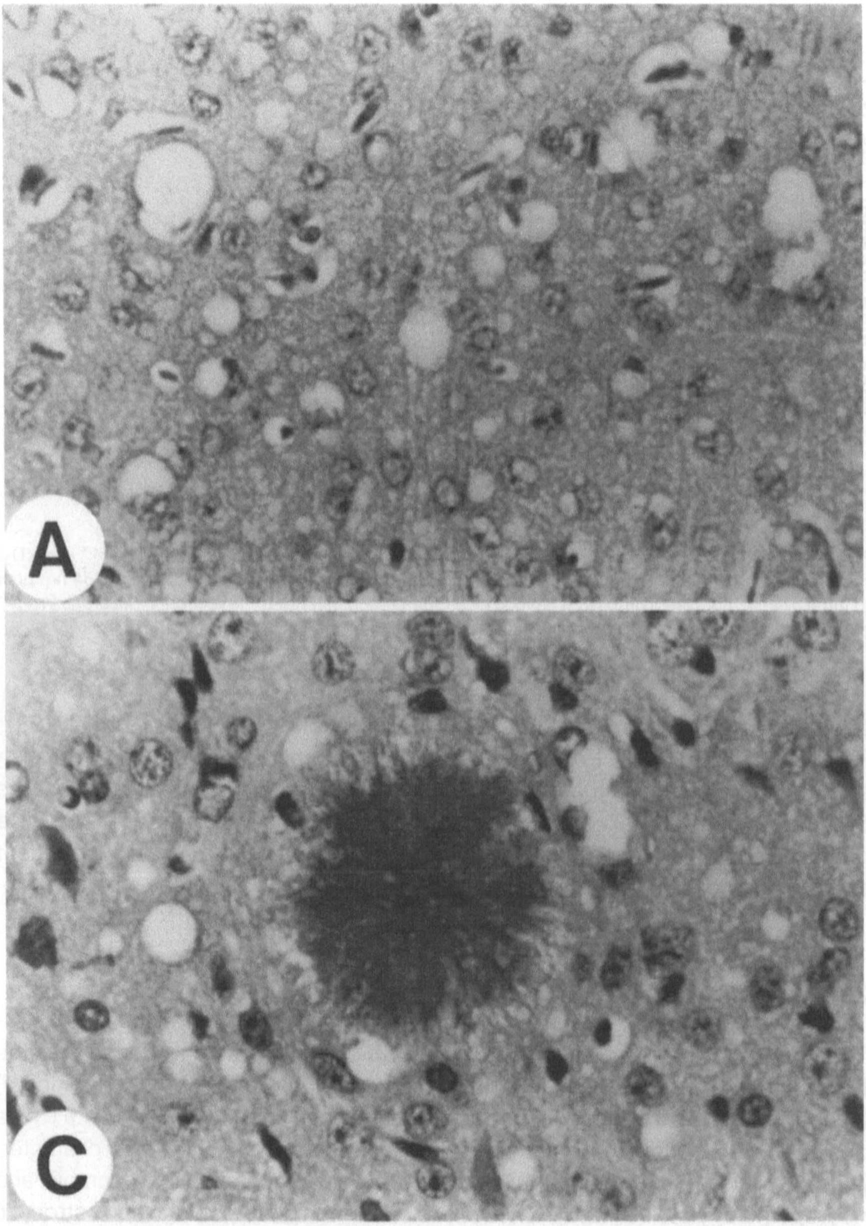

Fig. 3A–D. Neuropathology of Tg(MoPrP-P101L) mice developing neurodegeneration spontaneously. **A** Vacuolation in cerebral cortex of a Swiss CD-1 mouse that exhibited signs of neurologic dysfunction at 138 days after intracerebral inoculation with approx. 10^6 ID_{50} U RML scrapie prions. **B** Vacuolation in cerebral cortex of a Tg(MoPrP-P101L) mouse that exhibited signs of neurologic dysfunction at 252 days of age. **C** Kuru-type PrP amyloid plaque stained with periodic acid–Schiff in the caudate nucleus of a Tg(MoPrP-P101L) mouse that exhibited signs of neurologic dysfunction. **D** PrP amyloid plaques stained with α-PrP antiserum (RO73) in the caudate nucleus of a Tg(MoPrP-P101L) mouse that exhibited signs of neurologic dysfunction. *Bar* **(B)**, 50 μm (scale also applies to (**A**) *Bar* (**D**) 25 μm (scale also applies to **C**) Reprinted from PRUSINER 1993)

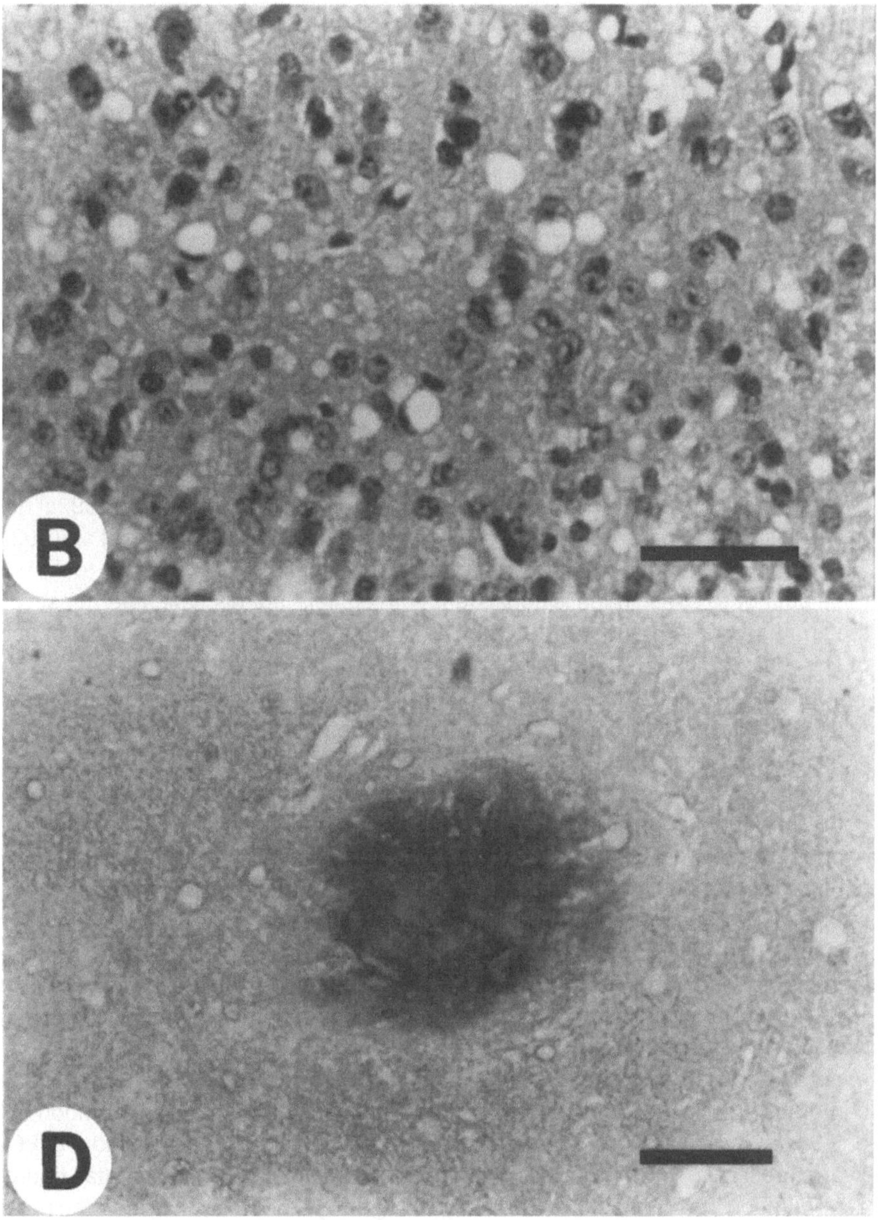

Fig. 3B,D

700 days after inoculation, while inoculated CD-1 Swiss mice remained well. Serial transmission of CNS degeneration in Tg196 mice required about 1 year while serial transmission in Syrian hamsters occurred after about 75 days (HSIAO et al. 1994). Although brain extracts prepared from Tg(MoPrP-P101L)H mice transmitted CNS degeneration to some inoculated recipients, little or no PrP^{Sc} was detected by immunoassays after limited proteolysis. Undetectable or low levels of PrP^{Sc} in the brains of these Tg(MoPrP-P101L)H mice are consistent with the results of these transmission experiments which suggest low titers of infectious prions. Though no PrP^{Sc} was detected in the brains of inoculated Tg196 mice exhibiting neurologic dysfunction by immunoassays after limited proteolysis, PrP amyloid plaques as well as spongiform degeneration were frequently found. The neurodegeneration found in inoculated Tg196 mice seems likely to result from a modification of mutant PrP^{C} that is initiated by mutant PrP^{Sc} present in the brain extracts prepared from ill Tg(MoPrP-P101L)H mice. In support of this explanation are the findings in some of the inherited human prion diseases as described above where neither protease-resistant PrP (BROWN et al. 1992; MEDORI et al. 1992a) nor transmission to experimental rodents could be demonstrated (TATEISHI et al. 1992). Furthermore, transmission of disease from Tg(MoPrP-P101L)H mice to Tg196 mice but not to Swiss mice is consistent with earlier findings which demonstrate that homotypic interactions between PrP^{C} and PrP^{Sc} feature in the formation of PrP^{Sc}.

In other studies, modifying the expression of mutant and wtPrP genes in Tg mice permitted experimental manipulation of the pathogenesis of both inherited and infectious prion diseases. Although overexpression of the wtPrP-A transgene approx. eight-fold was not deleterious to the mice, it did shorten scrapie incubation times from about 145 to about 45 days after inoculation with Mo scrapie prions (G.C. TELLING, T. HAGA, S.J. DEARMOND, and S.B. PRUSINER, in preparation). In contrast, overexpression at the same level of a PrP-A transgene mutated at codon 101 produced spontaneous, fatal neurodegeneration between 150 and 300 days of age in two new lines of Tg(MoPrP-P101L) mice designated 2866 and 2247. Genetic crosses of Tg(MoPrP-P101L) L2866 mice with gene targeted mice lacking both PrP alleles ($Prn\text{-}p^{0/0}$) produced animals with a highly synchronous onset of illness between 150 and 160 days of age. The Tg(MoPrP-P101L)2866/$Prn\text{-}p^{0/0}$ mice had numerous PrP plaques and widespread spongiform degeneration in contrast to the Tg2866 and 2247 mice that exhibited spongiform degeneration but only a few PrP amyloid plaques. Another line of mice designated Tg2862 overexpress the mutant transgene approx. 32-fold and develop fatal neurodegeneration between 200 and 400 days of age. Tg2862 mice exhibited the most severe spongiform degeneration and had numerous, large PrP amyloid plaques. While mutant PrP^{C}(P101L) clearly produces neurodegeneration, wtPrP^{C} profoundly modifies both the age of onset of illness and the neuropathology for a given level of transgene expression. These findings and those from other studies (TELLING et al. 1994) suggest that mutant and wtPrP interact, perhaps through a chaperonelike protein, to modify the pathogenesis of the dominantly inherited prion diseases.

3 Transmission of Prions

For three decades, the transmission of human prion diseases was studied largely with apes and monkeys where >90% of cases are thought to be transmissible (Brown et al. 1994; Gajdusek et al. 1966). Inoculations of mice, rats and hamsters produced variable results (Manuelidis et al. 1978; Tateishi and Kitamoto 1995; Tateishi et al. 1983). In our experience, only ~10% of intracerebrally inoculated mice developed CNS dysfunction with incubation times of >500 days (Prusiner 1987; Telling et al. 1994). Since previous investigations had shown that the "species barrier" between mice and Syrian hamsters for the transmission of prions can be abrogated by expression of a SHaPrP transgene in mice (Scott et al. 1989), Tg mice expressing HuPrP were constructed. These Tg(HuPrP) mice expressed levels of $HuPrP^C$ that were 4–8 fold higher than those of endogenous $MoPrP^C$; yet, upon inoculation with Hu prions, they failed to develop CNS dysfunction more frequently than non-Tg controls (Telling et al. 1994).

Because of the resistance of Tg(HuPrP) mice to Hu prions, we constructed mice expressing a chimeric Hu/Mo PrP transgene designated MHu2M. Earlier studies had shown that chimeric SHa/Mo PrP transgenes supported transmission of either Mo or SHa prions (Scott et al. 1993; Scott et al. 1992). The Tg(MHu2M) mice expressing the chimeric transgene at a level slightly below that of endogenous $MoPrP^C$ were found to be highly susceptible to Hu prions suggesting that Tg(HuPrP) mice have considerable difficulty converting $HuPrP^C$ into PrP^{Sc} (Telling et al. 1994). Although MoPrP and HuPrP differ at 28 residues, only nine or perhaps fewer amino acids in the region between codons 96 and 167 feature in the species barrier in the transmission of Hu prions into mice, as demonstrated by the susceptibility of Tg(MHu2M) mice to Hu prions.

To explore why Hu prions transmit disease to Tg(MHu2M) mice expressing chimeric PrP but not to Tg(HuPrP) mice, we crossed the Tg(HuPrP)FVB mice with those in which the MoPrP gene had been ablated, designated $Prnp^{0/0}$ (Büeler et al. 1992). The resulting Tg(HuPrP)$Prnp^{0/0}$ mice were found to be susceptible to Hu prions, whereas Tg(MHu2M) $Prnp^{0/0}$ mice were rendered only slightly more susceptible (Telling et al., In press). These observations contend that $MoPrP^C$ inhibited the conversion of $HuPrP^C$ into PrP^{Sc}; once $MoPrP^C$ was removed by gene ablation, inhibition was abolished.

3.1 Protein X and Prion Propagation

The results of our studies suggest that two separate domains of $HuPrP^C$ participate in the formation of PrP^{Sc}: (1) the central domain delimited by codons 96 and 167 as defined by the Hu sequence in chimeric $MHu2MPrP^C$ that binds to PrP^{Sc} and (2) an additional domain through which $HuPrP^C$ binds to a macromolecule other than PrP^{Sc} (Telling et al., In press). We assume that this macromolecule is a protein and have provisionally designated it "protein X." From our chimeric transgene studies, the second domain of PrP^C must be at the N- or C-terminus, i.e., outside the central region of PrP. Like the binding of PrP^C to PrP^{Sc} which is most efficient when the two isoforms have the same sequence (PRUSINER et al. 1990), the binding of PrP^C to protein X seems to exhibit the highest affinity when these two proteins are from the same species. Although the level of $MoPrP^C$ is only 10–20% of the transgene product $HuPrP^C$ in the brains of the Tg(HuPrP) mice, it prevented the conversion of $HuPrP^C$ into PrP^{Sc}. These findings suggest that $MoPrP^C$ binds to Mo protein X with a considerably higher affinity than does $HuPrP^C$, which provides an explanation for why $MoPrP^C$ inhibits the transmission of Hu prions in Tg(HuPrP) mice.

Since truncation of the N-terminus of recombinant PrP expressed in cultured cells still permitted the formation of PrP^{Sc}-like molecules (ROGERS et al. 1993), it seems likely that the site at which PrP^C binds to another protein is at the C-terminal end of PrP^C. Mature HuPrP differs from MoPrP at only 5 positions at the C-terminus which lie between residues 215 and 230, some of which are likely to form the protein X binding site for PrP^C (TELLING et al., In press).

The proposed model for prion propagation involving protein X is supported by studies on an inherited form of prion disease modeled in mice. Spontaneous CNS disease was found in uninoculated mice expressing the P102L point mutation of GSS when this substitution was introduced into MoPrP (HSIAO et al. 1994). The P102L mutation expressed in chimeric MHu2MPrP but not HuPrP produced CNS dysfunction in Tg mice. These findings argue that inherited prion disease like the transmissible disorder requires protein X for the conversion of mutant PrP^C into a pathologic isoform.

3.2 Protein Y and the Neuropathology of Prion Disease

Four lines of congenic mice were produced by crossing the PrP gene of the ILn/J mouse onto C57BL. The four lines of congenic mice are designated: B6.I-4 for B6.I-*B2m*[a], B6.I-1 for B6.I-*Prn-p*[b], B6.I-2 for B6.I-*Il-1a*[d] *Prn-p*[b], B6.I-3 for B6.I-*B2m*[a] *Prn-p*[b] (CARLSON et al. 1993). Neuropathologic examination of B6.I-1, B6.I-2, I/LnJ and VM/Dk mice inoculated with 87V prions showed numerous PrP amyloid plaques in accord with an earlier report on VM/Dk mice (BRUCE et al. 1976). In B6.I-1 mice intense spongiform degeneration, gliosis, and PrP immunostaining were found in the ventral posterior lateral nucleus of the thalamus, the habenula and the raphe nuclei of the brainstem (CARLSON et al. 1994). These same regions showed intense immunoreactivity for PrP^{Sc} on histoblots. Unexpectedly, B6.I-2 and ILn/J mice exhibited only mild vacuolation of the thalamus and brainstem. These findings suggest that a locus near *Prn-p* influences the deposition of PrP^{Sc}, and thus vacuolation, in the thalamus, the habenula and raphe nuclei. We have provisionally designated the product of this gene protein Y.

Identification of the gene that encodes protein Y that is distinct from but near *Prn-p* will be important. The gene Y product appears to control, at least in part, neuronal vacuolation and presumably PrP^{Sc} deposition in mice inoculated with scrapie prions. Isolation of protein Y should be helpful in dissecting the molecular events that feature in the pathogenesis of the prion diseases.

4 Prion Diversity

4.1 Prion Strains and Variations in Patterns of Disease

For many years studies of experimental scrapie were performed exclusively with sheep and goats. The disease was first transmitted by intraocular inoculation (Cuillé and Chelle 1939) and later by intracerebral, oral, subcutaneous, intramuscular, and intravenous injections of brain extracts from sheep developing scrapie. Incubation periods of 1–3 years were common, and often many of the inoculated animals failed to develop disease (Dickinson and Stamp 1969; Hadlow et al. 1980, 1982). Different breeds of sheep exhibited markedly different susceptibilities to scrapie prions inoculated subcutaneously, suggesting that the genetic background might influence host permissiveness (Gordon 1966).

The diversity of scrapie prions was first appreciated in goats inoculated with "hyper" and "drowsy" isolates (Pattison and Millson 1961). Subsequently, studies in mice demonstrated the existence of many scrapie "strains" (Bruce and Dickinson 1987; Dickinson and Fraser 1979; Dickinson and Outram 1988; Kimberlin et al. 1987) which continues to pose a fascinating conundrum. What is the macromolecule that carries the information required for each strain to manifest a unique set of biological properties if it is not a nucleic acid?

There is good evidence for multiple "strains" or distinct isolates of prions as defined by specific incubation times, distribution of vacuolar lesions and patterns of PrP^{Sc} accumulation (Bruce et al. 1989; Dickinson et al. 1968; Fraser and Dickinson 1973; Hecker et al. 1992). The lengths of the incubation times have been used to distinguish prion strains inoculated into sheep, goats, mice, and hamsters. Dickinson and his colleagues developed a system for "strain typing" by which mice with genetically determined short and long incubation times were used in combination with the F1 cross (Dickinson et al. 1968, 1984; Dickinson and Meikle 1971). For example, C57BL mice exhibited short incubation times of approx. 150 days when inoculated with either the Me7 or Chandler isolates; VM mice inoculated with these same isolates had prolonged incubation times of approx. 300 days. The mouse gene controlling incubation times was labeled *Sinc*, and long incubation times were said to be a dominant trait because of prolonged incubation times in F1 mice. Prion strains were categorized into two groups based upon their incubation times: (a) those causing disease more rapidly in "short" incubation time C57BL mice and (b) those causing disease more rapidly in "long" incubation time VM mice.

4.2 PrP Gene Dosage Controls the Length of Scrapie Incubation Time

More than a decade of study was required to unravel the mechanism responsible for the "dominance" of long incubation times; not unexpectedly, long incubation times were found not to be dominant traits. Instead, the apparent dominance of long incubation times is due to a gene dosage effect (CARLSON et al. 1994).

Our own studies began with the identification of a widely available mouse strain with long incubation times. ILn/J mice inoculated with Rocky Mountain Lab (RML) prions were found to have incubation times exceeding 200 days (KINGSBURY et al. 1983), a finding that was confirmed by others (CARP et al. 1987). Once molecular clones of the PrP gene were available, we asked whether the PrP genes of short and long mice segregate with incubation times. A restriction fragment length polymorphism of the PrP gene was used to follow the segregation of MoPrP genes (*Prn-p*) from short NZW or C57BL mice with long ILn/J mice in F1 and F2 crosses. This approach permitted the demonstration of genetic linkage between a *Prn-p* and a gene modulating incubation times (*Prn-i*) (CARLSON et al. 1986). Other investigators have confirmed the genetic linkage, and one group has shown that the incubation time gene *Sinc* is also linked to PrP (HUNTER et al. 1987; RACE et al. 1990). It now seems likely that the genes for PrP, *Prn-i*, and *Sinc* are all congruent; the term *Sinc* is no longer used (O'BRIEN 1993). The PrP sequences of NZW with short and long scrapie incubation times, respectively, differ at codons 108 (L →F) and 189 (T →V) (WESTAWAY et al. 1987).

Although the amino acid substitutions in PrP that distinguish *Prn-p^a* from *Prn-p^b* mice argued for the congruency of *Prn-p* and *Prn-i*, experiments with *Prn-p^a* mice expressing *Prn-p^b* transgenes demonstrated a "paradoxical" shortening of incubation times (WESTAWAY et al. 1991). We had predicted that these Tg mice would exhibit a prolongation of the incubation time after inoculation with RML prions based on (*Prn-p^a* x *Prn-p^b*) F1 mice which do exhibit long incubation times. We described these findings as "paradoxical shortening" because we and others had believed for many years that long incubation times are dominant traits (CARLSON et al. 1986; DICKINSON et al. 1968). From studies of congenic and transgenic mice expressing different numbers of the *a* and *b* alleles of *Prn-p* (Table 3) we now realize that these findings were not paradoxical; indeed, they result from increased PrP gene dosage (CARLSON et al. 1994). When the RML isolate was inoculated into congenic and transgenic mice, increasing the number of copies of the *a* allele was found to be the major determinant in reducing the incubation time; however, increasing the number of copies of the *b* allele also reduced the incubation time, but not to the same extent as that seen with the *a* allele (Table 3).

The discovery that incubation times are controlled by the relative dosage of *Prn-p^a* and *Prn-p^b* alleles was foreshadowed by studies of Tg(SHaPrP) mice in which the length of the incubation time after inoculation with SHa prions was inversely proportional to the transgene product SHaPrP^C (PRUSINER et al. 1990). Not

Table 3. MoPrP-A expression is a major determinant of incubation times in mice inoculated with the RML scrapie prions

Mice	PPrn-p Genotype (copies)	Prn-p Transgenes (copies)	Alleles *a*	Alleles *b*	Incubation Time[a] (days)	*n*
Prn-p$^{0/0}$	0/0		0	0	>600	4
Prn-p$^{+/0}$	a/0		1	0	426 ± 18	9[a]
B6.I–1	b/b		0	2	360 ± 16	7[b]
B6.I–2	b/b		0	2	379 ± 8	10[b]
B6.I–3	b/b		0	2	404 ± 10	20
(B6 x B6.I–1)F1	a/b		1	1	268 ± 4	7
B6.I–1 x Tg(MoPrP-B$^{0/0}$)15	a/b		1	1	255 ± 7	11[c]
B6.I–1 x Tg(MoPrP-B$^{0/0}$)15	a/b		1	1	274 ± 3	9[d]
B6.I–1 x Tg(MoPrP-B$^{+/0}$)15	a/b	bbb/0	1	4	166 ± 2	11[c]
B6.I–1 x Tg(MoPrP-B$^{+/0}$)15	a/b	bbb/0	1	4	162 ± 3	8[d]
C57BL/6J (B6)	a/a		2	0	143 ± 4	8
B6.I–4	a/a		2	0	144 ± 5	8
non-Tg(MoPrP-B$^{0/0}$)15	a/a		2	0	130 ± 3	10
Tg(MoPrP-B$^{+/0}$)15	a/a	bbb/0	2	3	115 ± 2	18
Tg(MoPrP-B$^{+/+}$)15	a/a	bbb/bbb	2	6	111 ± 5	5
Tg(MoPrP-B$^{+/0}$) 94	a/a	>30b	2	>30	75 ± 2	15[e]
Tg(MoPrP-A$^{+/0}$)B4053	a/a	>30a	>30	0	50 ±	16

[a] Data from PRUSINER et al. (1993b).
[b] Data from CARLSON et al. (1993).
[c] The homozygous Tg(MoPrP-B$^{+/+}$)15 mice were maintained as a distinct subline selected for transgene homozygosity two generations removed from the (B6 x LT/Sv)F2 founder. Hemizygous Tg(MoPrP-B$^{+/0}$)15 mice were produced by crossing the Tg(MoPrP-B$^{+/+}$)15 line with B6 mice.
[d] Tg(MoPrP-B$^{+/0}$)15 mice were maintained by repeated backcrossing to B6 mice.
[e] Data from WESTAWAY et al. (1991).

only does the PrP gene dose determine the length of the incubation time, but also the passage history of the inoculum, particularly in *Prn-p^b* mice (Table 4). The PrP^{Sc} allotype in the inoculum produced the shortest incubation times when it was the same as that of PrP^C in the host (CARLSON et al. 1989). The term "allotype" is used to describe allelic variants of PrP. To address the issue of whether gene products other than PrP might be responsible for these findings, we inoculated B6 and B6.1-4 mice carrying *Prn-p$^{a/a}$* as well as I/Ln, and B6.I-2 mice (CARLSON et al. 1993, 1994), with RML prions passaged in mice homozygous for either the *a* or *b* allele of *Prn-p* (Table 4). CD-1 and NZW/LacJ mice produce prions containing PrP^{Sc}-A encoded by *Prn-p^a* while I/LnJ mice produce PrP^{Sc}-B prions. The incubation times in the congenic mice reflected the PrP allotype, rather than other factors acquired during prion passage. The effect of the allotype barrier was small when measured in *Prn-p$^{a/a}$* mice but was clearly demonstrable in *Prn-p$^{b/b}$* mice. B6.I-2 congenic mice inoculated with prions from I/Ln mice had an incubation time of 237 ± 8 days compared to times of 360 ± 16 days and 404 ± 4 days for mice inoculated with prions passaged in CD-1 and NZW mice, respectively. Thus, previous passage of prions in *Prn-p^b* mice shortened the incubation time by approx. 40% when assayed in *Prn-p^b* mice, compared to those inoculated with prions passaged in *Prn-p^a* mice (CARLSON et al. 1989).

Table 4. Mismatching of PrP allotypes between PrP^{Sc} in the inoculum and PrP^{C} in the inoculated host extends prion incubation times in congenic mice

Mice	Host Genotype	Donor	Inoculum (genotype)	Donor Incubation time	*n*
C57BL/6J (B6)	a/a	CD-1	a/a	143 ± 4	8
B6.I–4	a/a	NZW	a/a	144 ± 5	8
B6.I–4	a/a	I/Ln	b/b	150 ± 6	6
B6.I–2	b/b	CD-1	a/a	360 ± 16	8
B6.I–2	b/b	NZW	a/a	404 ± 4	20
B6.I–2	b/b	I/Ln	b/b	237 ± 8	17
I/LnJ[a]	b/b	CD-1	a/a	314 ± 13	11
I/LnJ	b/b	NZW	a/a	283 ± 21	8
I/LnJ	b/b	I/Ln	b/b	193 ± 6	16

[a] I/LnJ results previously reported Carlson et al. (1994).

4.3 Overdominance

The phenomenon of "overdominance" in which incubation times in F1 hybrids are longer than those of either parent (Dickinson and Meikle 1969) contributed to the confusion surrounding control of scrapie incubation times. When the 22A scrapie isolate was inoculated into B6, B6.I-1 and (B6 x B6.I-1)F1, overdominance was observed: the scrapie incubation time in B6 mice was 405 ± 2 days, in B6.I mice 194 ± 10 days and in (B6 × B6.I-1)F1 mice 508 ± 14 days (Table 5). Shorter incubation times were observed in Tg(MoPrP-B)15 mice which were either homozygous or hemizygous for the *Prn-p*^{b} transgene. Hemizygous Tg(MoPrP-$B^{+/0}$)15 mice exhibited a scrapie incubation time of 395 ± 12 days while the homozygous mice had an incubation time of 286 ± 15 days.

As with the results with the RML isolate (Table 4), the findings with the 22A isolate can be explained on the basis of gene dosage; however, the relative effects of the *a* and *b* alleles differ in two respects. First, the *b* allele is the major determinant of the scrapie incubation time with the 22A isolate, not the *a* allele. Second, increasing the number of copies of the *a* allele does not diminish the incubation but prolongs it: the *a* allele is inhibitory with the 22A isolate (Table 5). With the 87V prion isolate the inhibitory effect of the *Prn-p*^{a} allele is even more pronounced since only a few *Prn-p*^{a} and (*Prn-p*^{a} × *Prn-p*^{b})F1 mice develop scrapie after more than 600 days postinoculation (Carlson et al. 1994).

The most interesting feature of the incubation time profile for 22A is the overdominance of the *a* allele of *Prn-p* in prolonging incubation period. On the basis of overdominance Dickinson and Outram (1979) put forth the replication site hypothesis postulating that dimers of the *Sinc* gene product feature in the replication of the scrapie agent. The results in Table 5 are compatible with the interpretation that the target for PrP^{Sc} may be a PrP^{C} dimer or multimer. The assumptions under this model are that PrP^{C}-B dimers are more readily converted to PrP^{Sc} than are PrP^{C}-A dimers and that PrP^{C}-A: PrP^{C}-B heterodimers are even more resistant to conversion to PrP^{Sc} than PrP^{C}-A dimers. Increasing the ratio of PrP-B to PrP-A would lead to shorter incubation times by favoring the formation of PrP^{C}-B homodimers (Table 5). A similar mechanism may account for the

Table 5. MoPrPA^{-C} inhibits the synthesis of 22A scrapie prions

Mice	Prn-p genotype	Prn-p transgenes (copies)	Alleles(copies)		Incubation time (days)	*n*
			a	*b*		
B6.I–1	b/b		0	2	194 ± 10	7
(B6 x B6.I–1)F1	a/b		1	1	508 ± 14	7
C57BL/6J (B6)	a/a		2	0	405 ± 2	8
non-Tg(MoPrP-B$^{0/0}$)15	a/a		2	0	378 ± 8	3[a]
Tg(MoPrP-B$^{+/0}$)15	a/a	bbb/0	2	3	318 ± 14	15[a]
Tg(MoPrP-B$^{+/0}$)15	a/a	bbb/0	2	3	395 ± 12	6[b]
Tg(MoPrP-B$^{+/+}$)15	a/a	bbb/bbb	2	6	266 ± 1	6[a]
Tg(MoPrP-B$^{+/+}$)15	a/a	bbb/bbb	2	6	286 ± 15	5[b]

[a] The homozygous Tg(MoPrP-B$^{+/+}$)15 mice were maintained as a distinct subline selected for transgene homozygosity two generations removed from the (B6 x LT/Sv)F2 founder. Hemizygous Tg(MoPrP-B$^{+/0}$)15 mice were produced by crossing the Tg(MoPrP-B$^{+/+}$)15 line with B6 mice.

[b] Tg(MoPrP-B$^{+/0}$)15 mice were maintained by repeated backcrossing to B6 mice.

relative paucity of individuals heterozygous for the Met/ Val polymorphism at codon 129 of the human PrP gene in spontaneous and iatrogenic CJD (Palmer et al. 1991). Alternatively, PrP^{C}-PrP^{Sc} interaction can be broken down to two distinct aspects: binding affinity and efficacy of conversion to PrP^{Sc}. If PrP-A has a higher affinity for 22A PrP^{Sc} than does PrP^{C}-B, but is inefficiently converted to PrP^{Sc}, the exceptionally long incubation time of *Prn-p$^{a/b}$* heterozygotes might reflect reduction in the supply of 22A prions available for interaction with the PrP^{C}-B product of the single *Prn-p^{b}* allele. Additionally, PrP^{C}-A may inhibit the interaction of 22A PrP^{Sc} with PrP^{C}-B leading to prolongation of the incubation time. This interpretation is supported by prolonged incubation times in Tg(SHaPrP) mice inoculated with mouse prions in which SHaPrP^{C} is thought to inhibit the binding of MoPrP^{Sc} to the substrate MoPrP^{C} (Prusiner et al. 1990).

4.4 Patterns of PrP^{Sc} Deposition

Besides measurements of the length of the incubation time, profiles of spongiform degeneration have also been used to characterize different prion strains (Fraser 1979; Fraser and Dickinson 1973). With the development of a new procedure for in situ detection of PrP^{Sc}, designated histoblotting (Taraboulos et al. 1992), it became possible to localize and quantify PrP^{Sc} as well as to determine whether "strains" produce different, reproducible patterns of PrP^{Sc} accumulation (DeArmond et al. 1993; Hecker et al. 1992).

Histoblotting overcame two obstacles that plagued PrP^{Sc} detection in brain by standard immunohistochemical techniques: the presence of PrP^{C} and weak antigenicity of PrP^{Sc} (DeArmond et al. 1987). The histoblot is made by pressing 10-µm-thick cryostat sections of fresh frozen brain tissue to nitrocellulose paper. To localize protease-resistant PrP^{Sc} in brain the histoblot is digested with proteinase K to eliminate PrP^{C}, followed by denaturation of the undigested PrP^{Sc} to enhance binding of PrP antibodies. Immunohistochemical staining yields a far more intense, specific and reproducible PrP signal than can be achieved by immunohistochemistry on standard tissue sections. The intensity of immunostaining correlates well with neurochemical estimates of PrP^{Sc} concentration in homogenates of dissected brain regions. PrP^{C} can be localized in histoblots of normal brains by eliminating the proteinase K digestion step.

Comparisons of PrP^{Sc} accumulation on histoblots with histologic sections showed that PrP^{Sc} deposition preceded vacuolation and only those regions with PrP^{Sc} underwent degeneration. Microdissection of individual brain regions confirmed the conclusions of the histoblot studies: those regions with high levels of PrP 27–30 had intense vacuolation (Casaccia-Bonnefil et al. 1993). Thus, we concluded that the deposition of PrP^{Sc} is responsible for the neuropathologic changes found in the prion diseases.

While studies with both mice and Syrian hamsters established that each isolate has a specific signature as defined by a specific pattern of PrP^{Sc} accumulation in the brain (Carlson et al. 1994; DeArmond et al. 1993; Hecker et al. 1992), comparisons must be done on an isogenic background (Hsiao et al. 1994; Scott et al. 1993). Variations in the patterns of PrP^{Sc} accumulation were found to be equally as great as those seen between two strains when a single strain is inoculated in mice expressing different PrP genes. Based upon the initial studies which were performed in animals of a single genotype we suggested that PrP^{Sc} synthesis occurs in specific populations of cells for a given distinct prion isolate.

4.5 Are Prion Strains Different PrP^{Sc} Conformers?

Explaining the problem of multiple distinct prion isolates might be accommodated by multiple PrP^{Sc} conformers that act as templates for the folding of de novo synthesized PrP^{Sc} molecules during prion "replication" (Fig. 4). Although it is clear that passage history can be responsible for the prolongation of incubation time when prions are passed between mice expressing different PrP allotypes (Carlson et al. 1989) or between species (Prusiner et al. 1990), many scrapie strains show distinct incubation times in the same inbred host (Bruce et al. 1991).

In recent studies we inoculated three strains of prions into congenic and Tg mice harboring various numbers of the *a* and *b* alleles of *Prn-p* (Carlson et al. 1994). The number of *Prn-p*^{a} genes was the major determinant of incubation times in mice inoculated with the RML prion isolate and was inversely related to the length of the incubation time (Table 3). In contrast, the *Prn-p*^{a} allele prevented scrapie in mice inoculated with 87V prions. *Prn-p*^{b} genes were permissive for 87V prions and shortened incubation times in most mice inoculated with 22A prions (Table 5). Experiments with the 87V isolate suggest that a genetic locus encoding protein Y, distinct from *Prn-p*, controls the deposition of PrP^{Sc} and the attendant neuropathology. While each prion isolate produced distinguishable patterns of PrP^{Sc} accumulation in brain, a comparison of these patterns showed that those patterns found with RML and 22A prions in congenic *Prn-p*^{b} mice were more similar than those with RML prions in *Prn-p*^{a} and *Prn-p*^{b} congenic mice. Thus, both the PrP genotype and prion isolate modify the distribution of PrP^{Sc} and the length of the incubation time. These findings suggest that prion strain specified properties result from different affinities of PrP^{Sc} in the inocula for PrP^C-A and PrP^C-B allotypes encoded by the host.

Although the proposal for multiple PrP^{Sc} conformers is rather unorthodox, we already know that PrP can assume at least two profoundly different conformations: PrP^C and PrP^{Sc} (PAN et al. 1993). Of note, two different isolates from mink dying of transmissible mink encephalopathy exhibit different sensitivities of PrP^{Sc} to proteolytic digestion, supporting the suggestion that isolate-specific information might be carried by PrP^{Sc} (BESSEN and MARSH 1992a,b; MARSH et al. 1991). How many conformations PrP^{Sc} can assume is unknown. The molecular weight of a PrP^{Sc} homodimer is consistent with the ionizing radiation target size of 55 ± 9 kDa as determined for infectious prion particles independent of their polymeric form

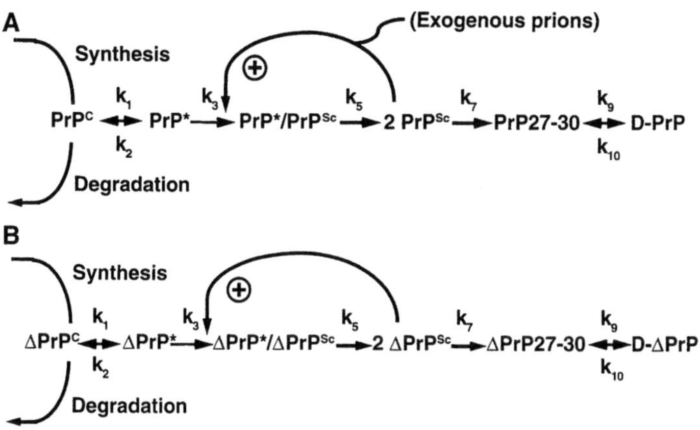

Fig. 4 A,B. Models for the replication of prions. **A** Proposed scheme for the replication of prions in sporadic and infectious prion diseases. wtPrP^C is synthesized and degraded as part of the normal metabolism of many cells. Stochastic fluctuations in the structure of PrP^C can create (k_1) a rare, partially unfolded, monomeric structure, PrP^*, that is an intermediate in the formation of PrP^{Sc}, but can revert (k_2) to PrP^C or be degraded prior to its conversion (k_3) into PrP^{Sc}. Normally the concentration of PrP^* is small and PrP^{Sc} formation is insignificant. In infectious prion diseases, exogenous prions enter the cell and stimulate conversion of PrP^* into PrP^{Sc}. In the absence of exogenous prions the concentration of PrP^{Sc} may eventually reach a threshold level in sporadic prion diseases after which a positive feedback loop would stimulate the formation of PrP^{Sc}. Limited proteolysis of the N-terminus of PrP^{Sc} produces (k_5) PrP 27–30 which can also be generated in scrapie-infected cells from a recombinant vector encoding PrP truncated at the N-terminus (ROGERS et al. 1993). Denaturation (k_9) of PrP^{Sc} or PrP 27–30 renders these molecules protease sensitive and abolishes scrapie infectivity; attempts to renature (k_{10}) these PrP^{Sc} or PrP 27–30 have been unsuccessful to date (PRUSINER et al. 1983, 1993c). **B** Scheme for the replication of prions in genetic prion diseases. Mutant (Δ) PrP^C is synthesized and degraded as part of the normal metabolism of many cells. Stochastic fluctuations in the structure of Δ PrP^C are increased compared to wtPrP^C, which creates (k_1) a partially unfolded, monomeric structure, ΔPrP^*, that is an intermediate in the formation of ΔPrP^{Sc}, but can revert (k_2) to ΔPrP^C or be degraded prior to its conversion (k_3) into ΔPrP^{Sc}. Limited proteolysis of the N-terminus of ΔPrP^{Sc} produces (k_5) ΔPrP 27–30 which in some cases may be less protease resistant than wtPrP 27–30 (MONARI et al. 1994; PRUSINER and HSIAO 1994). (Adapted from COHEN et al. 1994)

(Bellinger-Kawahara et al. 1988). If prions are oligomers of PrP^{Sc}, which seems likely, then this offers another level of complexity which in turn generates additional diversity.

5 Conclusion

5.1 Prions are not Viruses

The study of prions has taken several unexpected directions over the past few years. The discovery that prion diseases in humans are uniquely both genetic and infectious has greatly strengthened and extended the prion concept. To date 18 different mutations in the human PrP gene all resulting in nonconservative substitutions have been found to either be linked genetically to or segregate with the inherited prion diseases. However, the transmissible prion particle is composed largely, if not entirely, of an abnormal isoform of the prion protein designated PrP^{Sc} (Prusiner 1991). These findings argue that prion diseases should be considered pseudoinfections, since the particles transmitting disease appear to be devoid of a foreign nucleic acid and thus differ from all known microorganisms as well as viruses and viroids. Because much information, especially about scrapie of rodents, has been derived using experimental protocols adapted from virology, we continue to use terms such as infection, incubation period, transmissibility, and endpoint titration in studies of prion diseases.

5.2 Do Prions Exist in Lower Organisms?

In *Saccharomyces cervisiae*, ure2 and [URA3] mutants were described that can grow on ureidosuccinate under conditions of nitrogen repression such as glutamic acid and ammonia (Lacroute 1971). Mutants of URE2 exhibit Mendelian inheritance, whereas [URE3] is cytoplasmically inherited (Wickner 1994). The [URE3] phenotype can be induced by UV irradiation and by overexpression of ure2p, the gene product of ure2; deletion of ure2 abolishes [URE3]. The function of ure2p is unknown, but it has substantial homology with glutathione *S*-transferase; attempts to demonstrate this enzymic activity with purified ure2p have not been successful (Coschigano and Magasanik 1991). Whether the [URE3] protein is a posttranslationally modified form of ure2p which acts upon unmodified ure2p to produce more of itself remains to be established.

Another possible yeast prion is the [PSI] phenotype (Wickner 1994). [PSI] is a non-Mendelian inherited trait that can be induced by expression of the PNM2 gene (Cox et al. 1988). Both [PSI] and [URE3] can be cured by exposure of the yeast to 3 mM GdnHCl. The mechanism responsible for abolishing [PSI] and [URE3] with a low concentration of GdnHCl is unknown. In the filamentous fungus *Podospora anserina*, the het-s locus controls the vegetative incompatibility;

conversion from the S^s to the s state seems to be a posttranslational, autocatalytic process (DELEU et al. 1993).

If any of the above cited examples can be shown to function in a manner similar to prions in animals, many new, more rapid and economical approaches to prion diseases should be forthcoming.

5.3 Common Neurodegenerative Diseases

The knowledge accrued from the study of prion diseases may provide an effective strategy for defining the etiologies and dissecting the molecular pathogenesis of the more common neurodegenerative disorders such as Alzheimer's disease, Parkinson's disease, and amyotrophic lateral sclerosis. Advances in the molecular genetics of Alzheimer's disease and amyotrophic lateral sclerosis suggest that, as with the prion diseases, an important subset is caused by mutations that result in nonconservative amino acid substitutions in proteins expressed in the CNS (GOATE et al. 1991; LEVY et al. 1990; MULLAN et al. 1992; ROSEN et al. 1993; SCHELLENBERG et al. 1992; ST. GEORGE-HYSLOP et al. 1992; VAN BROECKHOVEN et al. 1990, 1992). Since people at risk for inherited prion diseases can now be identified decades before neurologic dysfunction is evident, the development of an effective therapy is imperative.

5.4 Future Studies

Tg mice expressing foreign or mutant PrP genes now permit virtually all facets of prion diseases to be studied and have created a framework for future investigations. Furthermore, the structure and organization of the PrP gene suggested that PrP^{Sc} is derived from PrP^{C} or a precursor by a posttranslational process. Studies with scrapie-infected cultured cells have provided much evidence that the conversion of PrP^{C} to PrP^{Sc} is a posttranslational process that probably occurs within a subcellular compartment bounded by cholesterol-rich membranes. The molecular mechanism of PrP^{Sc} formation remains to be elucidated, but chemical and physical studies have shown that the conformations of PrP^{C} and PrP^{Sc} are profoundly different.

The study of prion biology and diseases seems to be a new and emerging area of biomedical investigation. While prion biology has its roots in virology, neurology, and neuropathology, its relationships to the disciplines of molecular and cell biology as well as protein chemistry have become evident only recently. Certainly, the possibility that learning how prions multiply and cause disease will open up new vistas in biochemistry and genetics seems likely.

Acknowledgments. We thank M. Baldwin, D. Borchelt, G. Carlson, F. Cohen, C. Cooper, S. DeArmond, R. Fletterick, M. Gasset, R. Gabizon, D. Groth, R. Koehler, L. Hood, K. Hsiao, Z. Huang, V. Lingappa, K.-M. Pan, D. Riesner, A. Serban, N. Stahl, A. Taraboulos, M. Torchia, and D. Westaway for their help in these studies. Supported by grants from the National Institutes of Health (NS14069, AG08967,

AG02132, NS22786, and AG10770) and the American Health Assistance Foundation, as well as by gifts from Sherman Fairchild Foundation, Bernard Osher Foundation, and National Medical Enterprises.

References

Alper T, Haig DA, Clarke MC (1966) The exceptionally small size of the scrapie agent. Biochem Biophys Res Commun 22: 278–284

Alper T, Cramp WA, Haig DA, Clarke MC (1967) Does the agent of scrapie replicate without nucleic acid? Nature 214: 764–766

Alpers M (1987) Epidemiology and clinical aspects of kuru. In: Prusiner SB, McKinley MP (eds) Prions – novel infectious pathogens causing scrapie and Creutzfeldt-Jakob disease. Academic, Orlando, pp 451–465

Barry RA, Prusiner SB (1986) Monoclonal antibodies to the cellular and scrapie prion proteins. J Infect Dis 154: 518–521

Bellinger-Kawahara CG, Kempner E, Groth DF, Gabizon R, Prusiner SB (1988) Scrapie prion liposomes and rods exhibit target sizes of 55, 000 Da. Virology 164: 537–541

Bessen RA, Marsh RF (1992a) Biochemical and physical properties of the prion protein from two strains of the transmissible mink encephalopathy agent. J Virol 66: 2096–2101

Bessen RA, Marsh RF (1992b) Identification of two biologically distinct strains of transmissible mink encephalopathy in hamsters. J Gen Virol 73: 329–334

Bockman JM, Prusiner SB, Tateishi J, Kingsbury DT (1987) Immunoblotting of Creutzfeldt-Jakob disease prion proteins: host species-specific epitopes. Ann Neurol 21: 589–595

Borchelt DR, Scott M, Taraboulos A, Stahl N, Prusiner SB (1990) Scrapie and cellular prion proteins differ in their kinetics of synthesis and topology in cultured cells. J Cell Biol 110: 743–752

Brown P (1992) The phenotypic expression of different mutations in transmissible human spongiform encephalopathy. Rev Neurol 148: 317–327

Brown P, Cathala F, Raubertas RF, Gajdusek DC, Castaigne P (1987) The epidemiology of Creutzfeldt-Jakob disease: conclusion of 15-year investigation in France and review of the world literature. Neurology 37: 895–904

Brown P, Goldfarb LG, Kovanen J, Haltia M, Cathala F, Sulima M, Gibbs CJ Jr., Gajdusek DC (1992) Phenotypic characteristics of familial Creutzfeldt-Jakob disease associated with the codon 178^{Asn} *PRNP* mutation. Ann Neurol 31: 282–285

Brown P, Gibbs CJ Jr., Rodgers-Johnson P, Asher DM, Sulima MP, Bacote A, Goldfarb LG, Gajdusek DC (1994) Human spongiform encephalopathy: the National Institutes of Health series of 300 cases of experimentally transmitted disease. Ann Neurol 35: 513–529

Bruce ME, Dickinson AG (1987) Biological evidence that the scrapie agent has an independent genome. J Gen Virol 68: 79–89

Bruce ME, Dickinson AG, Fraser H (1976) Cerebral amyloidosis in scrapie in the mouse: effect of agent strain and mouse genotype. Neuropathol Appl Neurobiol 2: 471–478

Bruce ME, McBride PA, Farquhar CF (1989) Precise targeting of the pathology of the sialoglycoprotein, PrP, and vacuolar degeneration in mouse scrapie. Neurosci Lett 102: 1–6

Bruce ME, McConnell I, Fraser H, Dickinson AG (1991) The disease characteristics of different strains of scrapie in *Sinc* congenic mouse lines: implications for the nature of the agent and host control of pathogenesis. J Gen Virol 72: 595–603

Büeler H, Fischer M, Lang Y, Bluethmann H, Lipp H-P, DeArmond SJ, Prusiner SB, Aguet M, Weissmann C (1992) Normal development and behaviour of mice lacking the neuronal cell-surface PrP protein. Nature 356: 577–582

Büeler H, Aguzzi A, Sailer A, Greiner R-A, Autenried P, Aguet M, Weissmann C (1993) Mice devoid of PrP are resistant to scrapie. Cell 73: 1339–1347

Carlson GA, Kingsbury DT, Goodman PA, Coleman S, Marshall ST, DeArmond SJ, Westaway D, Prusiner SB (1986) Linkage of prion protein and scrapie incubation time genes. Cell 46: 503–511

Carlson GA, Westaway D, DeArmond SJ, Peterson-Torchia M, Prusiner SB (1989) Primary structure of prion protein may modify scrapie isolate properties. Proc Natl Acad Sci USA 86: 7475–7479

Carlson GA, Ebeling C, Torchia M, Westaway D, Prusiner SB (1993) Delimiting the location of the scrapie prion incubation time gene on chromosome 2 of the mouse. Genetics 133: 979–988

Carlson GA, Ebeling C, Yang S-L, Telling G, Torchia M, Groth D, Westaway D, DeArmond SJ, Prusiner SB (1994) Prion isolate specified allotypic interactions between the cellular and scrapie prion proteins in congenic and transgenic mice. Proc Natl Acad Sci USA 91: 5690–5694
Carp RI, Moretz RC, Natelli M, Dickinson AG (1987) Genetic control of scrapie: incubation period and plaque formation in mice. J Gen Virol 68: 401–407
Casaccia-Bonnefil P, Kascsak RJ, Fersko R, Callahan S, Carp RI (1993) Brain regional distribution of prion protein PrP27–30 in mice stereotaxically microinjected with different strains of scrapie. J Infect Dis 167: 7–12
Cohen FE, Pan K-M, Huang Z, Baldwin M, Fletterick RJ, Prusiner SB (1994) Structural clues to prion replication. Science 264: 530–531
Coschigano PW, Magasanik B (1991) The *URE2* gene product of Saccharomyces cerevisiae plays an important role in the cellular response to the nitrogen source and has homology to glutathione S-transferases. Mol Cell Biol 11: 822–832
Cousens SN, Harries-Jones R, Knight R, Will RG, Smith PG, Matthews WB (1990) Geographical distribution of cases of Creutzfeldt-Jakob disease in England and Wales 1970–84. J. Neurol Neurosurg Psychiatry 53: 459–465
Cox BS, Tuite MF, McLaughlin CS (1988) The psi factor of yeast: a problem in inheritance. Yeast 4: 159–178
Cuillé J, Chelle PL (1939) Experimental transmission of trembling to the goat. CR Seances Acad Sci 208: 1058–1060
DeArmond SJ, Mobley WC, DeMott DL, Barry RA, Beckstead JH, Prusiner SB (1987) Changes in the localization of brain prion proteins during scrapie infection. Neurology 37: 1271–1280
DeArmond SJ, Yang S-L, Lee A, Bowler R, Taraboulos A, Groth D, Prusiner SB (1993) Three scrapie prion isolates exhibit different accumulation patterns of the prion protein scrapie isoform. Proc Natl Acad Sci USA 90: 6449–6453
Deleu C, Clavé C, Bégueret J (1993) A single amino acid difference is sufficient to elicit vegetative incompatibility in the fungus Podospora anserina. Genetics 135: 45–52
Dickinson AG, Fraser H (1979) An assessment of the genetics of scrapie in sheep and mice. In: Prusiner SB, Hadlow WJ (eds) Slow transmissible diseases of the nervous system, vol 1. Academic, New York, pp 367–386
Dickinson AG, Meikle VM (1969) A comparison of some biological characteristics of the mouse-passaged scrapie agents, 22A and ME7. Genet Res 13: 213–225
Dickinson AG, Meikle VMH (1971) Host-genotype and agent effects in scrapie incubation: change in allelic interaction with different strains of agent. Mol Gen Genet 112: 73–79
Dickinson AG, Outram GW (1979) The scrapie replication-site hypothesis and its implications for pathogenesis. In: Prusiner SB, Hadlow WJ (eds) Slow transmissible diseases of the nervous system, vol 2. Academic, New York, pp 13–31
Dickinson AG, Outram GW (1988) Genetic aspects of unconventional virus infections: the basis of the virino hypothesis. In: Bock G, Marsh J (eds) Novel infectious agents and the central nervous system. Wiley, Chichester, UK, pp 63–83 (Ciba Foundation symposium 135)
Dickinson AG, Stamp JT (1969) Experimental scrapie in Cheviot and Suffolk sheep. J Comp Pathol 79: 23–26
Dickinson AG, Young GB, Stamp JT, Renwick CC (1965) An analysis of natural scrapie in Suffolk sheep. Heredity 20: 485–503
Dickinson AG, Meikle VMH, Fraser H (1968) Identification of a gene which controls the incubation period of some strains of scrapie agent in mice. J Comp Pathol 78: 293–299
Dickinson AG, Bruce ME, Outram GW, Kimberlin RH (1984) Scrapie strain differences: the implications of stability and mutation. In: Tateishi J (ed) Proceedings of the workshop on slow transmissible diseases. Japanese Ministry of Health and Welfare, Tokyo, pp 105–118
Fraser H (1979) Neuropathology of scrapie: the precision of the lesions and their diversity. In: Prusiner SB, Hadlow WJ (eds) Slow transmissible diseases of the nervous system, vol 1. Academic, New York, pp 387–406
Fraser H, Dickinson AG (1973) Scrapie in mice. Agent-strain differences in the distribution and intensity of grey matter vacuolation. J Comp Pathol 83: 29–40
Gajdusek DC (1977) Unconventional viruses and the origin and disappearance of kuru. Science 197: 943–960
Gajdusek DC (1985) Subacute spongiform virus encephalopathies caused by unconventional viruses. In: Maramorosch K, McKelvey JJ Jr (eds) Subviral pathogens of plants and animals: viroids and prions. Academic, Orlando, pp 483–544
Gajdusek DC, Zigas V (1957) Degenerative disease of the central nervous system in New Guinea - the endemic occurrence of "Kuru" in the native population. N Engl J Med 257: 974–978

Gajdusek DC, Zigas V (1959) Clinical, pathological and epidemiological study of an acute progressive degenerative disease of the central nervous system among natives of the eastern highlands of New Guinea. Am J Med 26: 442–469

Gajdusek DC, Gibbs CJ Jr, Alpers M (1966) Experimental transmission of a kuru-like syndrome to chimpanzees. Nature 209: 794–796

Goate A, Chartier-Harlin M-C, Mullan M, Brown J, Crawford F, Fidani L, Giuffra L, Haynes A, Irving N, James L, Mant R, Newton P, Rooke K, Roques P, Talbot C, Pericak-Vance M, Roses A, Williamson R, Rossor M, Owen M, Hardy J (1991) Segregation of a missense mutation in the amyloid precursor protein gene with familial Alzheimer's disease. Nature 349: 704–706

Goldmann W, Hunter N, Martin T, Dawson M, Hope J (1991) Different forms of the bovine PrP gene have five or six copies of a short, G-C-rich element within the protein-coding exon. J Gen Virol 72: 201–204

Gordon WS (1946) Advances in veterinary research. Vet Res 58: 516–520

Gordon WS (1966) Variation in susceptibility of sheep to scrapie and genetic implications. In: Report of Scrapie Seminar, ARS 91–53. US Department of Agriculture, Washington DC, pp 53–67

Hadlow WJ, Kennedy RC, Race RE, Eklund CM (1980) Virologic and neurohistologic findings in dairy goats affected with natural scrapie. Vet Pathol 17: 187–199

Hadlow WJ, Kennedy RC, Race RE (1982) Natural infection of Suffolk sheep with scrapie virus. J Infect Dis 146: 657–664

Harries-Jones R, Knight R, Will RG, Cousens S, Smith PG, Matthews WB (1988) Creutzfeldt-Jakob disease in England and Wales, 1980–1984: a case-control study of potential risk factors. J Neurol Neurosurg Psychiatry 51: 1113–1119

Hecker R, Taraboulos A, Scott M, Pan K-M, Torchia M, Jendroska K, DeArmond SJ, Prusiner SB (1992) Replication of distinct prion isolates is region specific in brains of transgenic mice and hamsters. Genes Dev 6: 1213–1228

Hope J, Reekie LJD, Hunter N, Multhaup G, Beyreuther K, White H, Scott AC, Stack MJ, Dawson M, Wells GAH (1988) Fibrils from brains of cows with new cattle disease contain scrapie-associated protein. Nature 336: 390–392

Hsiao K, Prusiner SB (1990) Inherited human prion diseases. Neurology 40: 1820–1827

Hsiao K, Baker HF, Crow TJ, Poulter M, Owen F, Terwilliger JD, Westaway D, Ott J, Prusiner SB (1989) Linkage of a prion protein missense variant to Gerstmann-sträussler syndrome. Nature 338: 342–345

Hsiao KK, Scott M, Foster D, Groth DF, DeArmond SJ, Prusiner SB (1990) Spontaneous neurodegeneration in transgenic mice with mutant prion protein. Science 250: 1587–1590

Hsiao KK, Groth D, Scott M, Yang S-L, Serban H, Rapp D, Foster D, Torchia M, DeArmond SJ, Prusiner SB (1994) Serial transmission in rodents of neurodegeneration from transgenic mice expressing mutant prion protein. Proc Natl Acad Sci USA 91: 9126–9130

Hunter GD (1972) Scrapie: a prototype slow infection. J infect Dis 125: 427–440

Hunter N, Hope J, McConnell I, Dickinson AG (1987) Linkage of the scrapie-associated fibril protein (PrP) gene and Sinc using congenic mice and restriction fragment length polymorphism analysis. J Gen Virol 68: 2711–2716

Kascsak RJ, Rubenstein R, Merz PA, Tonna-DeMasi M, Fersko R, Carp RI, Wisniewski HM, Diringer H (1987) Mouse polyclonal and monoclonal antibody to scrapie-associated fibril proteins. J Virol 61: 3688–3693

Kimberlin RH, Cole S, Walker CA (1987) Temporary and permanent modifications to a single strain of mouse scrapie on transmission to rats and hamsters. J Gen Virol 68: 1875–1881

Kingsbury DT, Kasper KC, Stites DP, Watson JD, Hogan RN, Prusiner SB (1983) Genetic control of scrapie and Creutzfeldt-Jakob disease in mice. J Immunol 131: 491–496

Kirschbaum WR (1924) Zwei eigenartige Erkrankungen des Zentralnervensystems nach Art der spastischen Pseudosklerose (Jakob). Z Ges Neurol Psychiatr 92: 175–220

Kretzschmar HA, Stowring LE, Westaway D, Stubblebine WH, Prusiner SB, DeArmond SJ (1986) Molecular cloning of a human prion protein cDNA. DNA 5: 315–324

Laçroute F (1971) Non-Mendelian mutation allowing ureidosuccinic acid uptake in yeast. J Bacteriol 106: 519–522

Levy E, Carman MD, Fernandez-Madrid IJ, Power MD, Lieberburg I, van Duinen SG, Bots GTAM, Luyendijk W, Frangione B (1990) Mutation of the Alzheimer's disease amyloid gene in hereditary cerebral hemorrhage, Dutch type. Science 248: 1124–1126

Malmgren R, Kurland L, Mokri B, Kurtzke J (1979) The epidemiology of Creutzfeldt-Jakob disease. In: Prusiner SB, Hadlow WJ (eds) Slow transmissible diseases of the nervous system, vol 1. Academic, New York, pp 93–112

Manuelidis E, Gorgacz EJ, Manuelidis L (1978) Interspecies transmission of Creutzfeldt-Jakob disease to Syrian hamsters with reference to clinical syndromes and strains of agent. Proc Natl Acad Sci USA 75: 3422–3436
Marsh RF, Bessen RA, Lehmann S, Hartsough GR (1991) Epidemiological and experimental studies on a new incident of transmissible mink encephalopathy. J Gen Virol 72: 589–594
Masters CL, Harris JO, Gajdusek DC, Gibbs CJ Jr., Bernouilli C, Asher DM (1978) Creutzfeldt-Jakob disease: patterns of worldwide occurrence and the significance of familial and sporadic clustering. Ann Neurol 5: 177–188
Masters CL, Gajdusek DC, Gibbs CJ Jr (1981) Creutzfeldt-Jakob disease virus isolations from the Gerstmann-Sträussler syndrome. Brain 104: 559–588
Medori R, Montagna P, Tritschler HJ, LeBlanc A, Cortelli P, Tinuper P, Lugaresi E, Gambetti P (1992a) Fatal familial insomnia: a second kindred with mutation of prion protein gene at codon 178: Neurology 42: 669–670
Medori R, Tritschler H-J, LeBlanc A, Villare F, Manetto V, Chen HY, Xue R, Leal S, Montagna P, Cortelli P, Tinuper P, Avoni P, Mochi M, Baruzzi A, Hauw JJ, Ott J, Lugaresi E, Autilio-Gambetti L, Gambetti P (1992b) Fatal familial insomnia, a prion disease with a mutation at codon 178 of the prion protein gene. N Engl J Med 326: 444–449
Meggendorfer F (1930) Klinische und genealogische Beobachtungen bei einem Fall von spastischer Pseudosklerose Jakobs. Z Ges Neurol Psychiatr 128: 337–341
Monari L, Chen SG, Brown P, Parchi P, Petersen RB, Mikol J, Gray F, Cortelli P, Montagna P, Ghetti B, Goldfarb LG, Gajdusek DC, Lugaresi E, Gambetti P, Autilio-Gambetti L (1994) Fatal familial insomnia and familial Creutzfeldt-Jakob disease: different prion proteins determined by a DNA polymorphism. Proc Natl Acad Sci USA 91: 2839–2842
Mullan M, Houlden H, Windelspecht M, Fidani L, Lombardi C, Diaz P, Rossor M, Crook R, Hardy J, Duff K, Crawford F (1992) A locus for familial early-onset Alzheimer's disease on the long arm of chromosome 14, proximal to the α 1-antichymotrypsin gene. Nat Genet 2: 340–342
O'Brien SJ (ed) (1993) Genetic maps-locus maps of complex genomes, 6th edn. Cold Spring Harbor Laboratory Press, Cold Spring Harbor, New York pp 4.42–4.45
Oesch B, Teplow DB, Stahl N, Serban D, Hood LE, Prusiner SB (1990) Identification of cellular proteins binding to the scrapie prion protein. Biochemistry 29: 5848–5855
Palmer MS, Dryden AJ, Hughes JT, Collinge J (1991) Homozygous prion protein genotype predisposes to sporadic Creutzfeldt-Jakob disease. Nature 352: 340–342
Pan K-M, Baldwin M, Nguyen J, Gasset M, Serban A, Groth D, Mehlhorn I, Huang Z, Fletterick RJ, Cohen FE, Prusiner SB (1993) Conversion of α-helices into β-sheets features in the formation of the scrapie prion proteins. Proc Natl Acad Sci USA 90: 10962–10966
Parry HB (1962) Scrapie: a transmissible and hereditary disease of sheep. Heredity 17: 75–105
Parry HB (1983) Scrapie disease in sheep. Academic, New York
Pattison IH (1965) Experiments with scrapie with special reference to the nature of the agent and the pathology of the disease. In: Gajdusek DC, Gibbs CJ Jr, Alpers MP (eds) Slow, latent and temperate virus infections. US Government Printing, Washington DC, pp 249–257 (NINDB monograph 2)
Pattison IH (1966) The relative susceptibility of sheep, goats and mice to two types of the goat scrapie agent. Res Vet Sci 7: 207–212
Pattison IH, Jones KM (1967) The possible nature of the transmissible agent of scrapie. Vet Rec 80: 1–8
Pattison IH, Millson GC (1961) Scrapie produced experimentally in goats with special reference to the clinical syndrome. J Comp Pathol 71: 101–108
Prusiner SB (1982) Novel proteinaceous infectious particles cause scrapie. Science 216: 136–144
Prusiner SB (1987) The biology of prion transmission and replication. In: Prusiner SB, McKinley MP (eds) Prions-Novel Infectious Pathogens Causing Scrapie and Creutzfeldt-Jakob Disease. Academic Press, Orlando, pp 83–112
Prusiner SB (1989) Scrapie prions. Annu Rev Microbiol 43: 345–374
Prusiner SB (1991) Molecular biology of prion diseases. Science 252: 1515–1522
Prusiner SB (1992) Molecular biology and genetics of neurodegenerative diseases caused by prions. Adv Virus Res 41: 241–280
Prusiner SB (1993) Transgenetics and cell biology of prion diseases: investigations of PrP^{Sc} synthesis and diversity. Br Med Bul 49: 873–912
Prusiner SB (1994) Inherited prion diseases. Proc Natl Acad Sci USA 91: 4611–4614
Prusiner SB, Hsiao KK (1994) Human prion diseases. Ann Neurol 35: 385–395
Prusiner SB, McKinley MP, Bowman KA, Bolton DC, Bendheim PE, Groth DF, Glenner GG (1983) Scrapie prions aggregate to form amyloid-like birefringent rods. Cell 35: 349–358

Prusiner SB, Scott M, Foster D, Pan K-M, Groth D, Mirenda C, Torchia M, Yang S-L, Serban D, Carlson GA, Hoppe PC, Westaway D, DeArmond SJ (1990) Transgenetic studies implicate interactions between homologous PrP isoforms in scrapie prion replication. Cell 63: 673–686

Prusiner SB, Fuzi M, Scott M, Serban D, Serban H, Taraboulos A, Gabriel J-M, Wells G, Wilesmith J, Bradley R, DeArmond SJ, Kristensson K (1993a) Immunologic and molecular biological studies of prion proteins in bovine spongiform encephalopathy. J Infect Dis 167: 602–613

Prusiner SB, Groth D, Serban A, Koehler R, Foster D, Torchia M, Burton D, Yang S-L, DeArmond SJ (1993b) Ablation of the prion protein (PrP) gene in mice prevents scrapie and facilitates production of anti-PrP antibodies. Proc Natl Acad Sci USA 90: 10608–10612

Prusiner SB, Groth D, Serban A, Stahl N, Gabizon R (1993c) Attempts to restore scrapie prion infectivity after exposure to protein denaturants. Proc Natl Acad Sci USA 90: 2793–2797

Race RE, Graham K, Ernst D, Caughey B, Chesebro B (1990) Analysis of linkage between scrapie incubation period and the prion protein gene in mice. J Gen Virol 71: 493–497

Raeber AJ, Borchelt DR, Scott M, Prusiner SB (1992) Attempts to convert the cellular prion protein into the scrapie isoform in cell-free systems. J Virol 66: 6155–6163

Riesner D, Kellings K, Meyer N, Mirenda C, Prusiner SB (1992) Nucleic acids and scrapie prions. In: Prusiner SB, Collinge J, Powell J, Anderton B (eds) Prion diseases of humans and animals. Ellis Horwood, London, pp 341–358

Rogers M, Serban D, Gyuris T, Scott M, Torchia T, Prusiner SB (1991) Epitope mapping of the Syrian hamster prion protein utilizing chimeric and mutant genes in a vaccinia virus expression system. J Immunol 147: 3568–3574

Rogers M, Yehiely F, Scott M, Prusiner SB (1993) Conversion of truncated and elongated prion proteins into the scrapie isoform in cultured cells. Proc Natl Acad Sci USA 90: 3182–3186

Rosen DR, Siddique T, Patterson D, Figiewicz DA, Sapp P, Hentati A, Donaldson D, Goto J, O'Regan JP, Deng H-, Rahmani Z, Krizus A, McKenna-Yasek D, Cayabyab A, Gaston SM, Berger R, Tanzi RE, Halperin JJ, Herzfeldt B, Van den Bergh R, Hung W-Y, Bird T, Deng G, Mulder DW, Smyth C, Laing NG, Soriano E, Pericak-Vance MA, Haines J, Rouleau GA, Gusella JS, Horvitz HR, Brown RH Jr (1993) Mutations in Cu/Zn superoxide dismutase gene are associated with familial amyotrophic lateral slerosis. Nature 362: 59–62

Schellenberg GD, Bird TD, Wijsman EM, Orr HT, Anderson L, Nemens E, White JA, Bonnycastle L, Weber JL, Alonso ME, Potter H, Heston LL, Martin GM (1992) Genetic linkage evidence for a familial Alzheimer's disease locus on chromosome 14. Science 258: 668–671

Scott M, Foster D, Mirenda C, Serban D, Coufal F, Wälchli M, Torchia M, Groth D, Carlson G, DeArmond SJ, Westaway D, Prusiner SB (1989) Transgenic mice expressing hamster prion protein produce species-specific scrapie infectivity and amyloid plaques. Cell 59: 847–857

Scott M, Groth D, Foster D, Torchia M, Yang S-L, DeArmond SJ, Prusiner SB (1993) Propagation of prions with artificial properties in transgenic mice expressing chimeric PrP genes. Cell 73: 979–988

Scott MR, Köhler R, Foster D, Prusiner SB (1992) Chimeric prion protein expression in cultured cells and transgenic mice. Protein Sci 1: 986–997

Sigurdsson B (1954) Rida, a chronic encephalitis of sheep with general remarks on infections which develop slowly and some of their special characteristics. Br Vet J 110: 341–354

Sparkes RS, Simon M, Cohn VH, Fournier REK, Lem J, Klisak I, Heinzmann C, Blatt C, Lucero M, Mohandas T, DeArmond SJ, Westaway D, Prusiner SB, Weiner LP (1986) Assignment of the human and mouse prion protein genes to homologous chromosomes. Proc Natl Acad Sci USA 83: 7358–7362

St. George-Hyslop P, Haines J, Rogaev E, Mortilla M, Vaula G, Pericak-Vance M, Foncin J-F, Montesi M, Bruni A, Sorbi S, Rainero I, Pinessi L, Pollen D, Polinsky R, Nee L, Kennedy J, Macciardi F, Rogaeva E, Liang Y, Alexandrova N, Lukiw W, Schlumpf K, Tanzi R, Tsuda T, Farrer L, Cantu J-M, Duara R, Amaducci L, Bergamini L, Gusella J, Roses A, McLachlan DC (1992) Genetic evidence for a novel familial Alzheimer's disease locus on chromosome 14. Nat Genet 2: 330–334

Stahl N, Borchelt DR, Hsiao K, Prusiner SB (1987) Scrapie prion protein contains a phosphatidylinositol glycolipid. Cell 51: 229–240

Stahl N, Baldwin MA, Teplow DB, Hood L, Gibson BW, Burlingame AL, Prusiner SB (1993) Structural analysis of the scrapie prion protein using mass spectrometry and amino acid sequencing. Biochemistry 32: 1991–2002

Taraboulos A, Jendroska K, Serban D, Yang S-L, DeArmond SJ, Prusiner SB (1992) Regional mapping of prion proteins in brains. Proc Natl Acad Sci USA 89: 7620–7624

Tateishi J, Kitamoto T (1995) Inherited prion diseases and transmission to rodents. Brain Pathol 5: 53–59

Tateishi J, Sato Y, Ohta M (1983) Creutzfeldt-Jakob disease in humans and laboratory animals. In: Zimmerman HM (eds) Progress in Neuropathology, Vol.5. Raven Press, New York, pp 195–221

Tateishi J, Doh-ura K, Kitamoto T, Tranchant C, Steinmetz G, Warter JM, Boellaard JW (1992) Prion protein gene analysis and transmission studies of Creutzfeldt-Jakob disease. In: Prusiner SB, Collinge J, Powell J, Anderton B (eds) Prion diseases of humans and animals. Ellis Horwood, London, pp 129–134

Telling GC, Scott M, Foster D, Yang S-L, Torchia M, Sidle KCL, Collinge J, DeArmond SJ, Prusiner SB (1994) Transmission of Creutzfeldt-Jakob disease from humans to transgenic mice expressing chimeric human-mouse prion protein. Proc Natl Acad Sci USA 91: 9936–9940

Telling GC, Scott M, Mastrianni J, Gabizon R, Torchia M, Cohen FE, DeArmond SJ, Prusiner SB (In press) Prion propagation in mice expressing human and chimeric PrP transgenes implicates the interaction of cellular PrP with another protein. Cell

Van Broeckhoven C, Haan J, Bakker E, Hardy JA, Van Hul W, Wehnert A, Vegter-Van der Vlis M, Roos RA (1990) Amyloid β protein precursor gene and hereditary cerebral hemorrhage with amyloidosis (Dutch). Science 248: 1120–1122

Van Broeckhoven C, Backhovens H, Cruts M, De Winter G, Bruyland M, Cras P, Martin J-J (1992) Mapping of a gene predisposing to early-onset Alzheimer's disease to chromosome 14q24.3. Nat Genet 2: 335–339

Westaway D, Goodman PA, Mirenda CA, McKinley MP, Carlson GA, Prusiner SB (1987) Distinct prion proteins in short and long scrapie incubation period mice. Cell 51: 651–662

Westaway D, Mirenda CA, Foster D, Zebarjadian Y, Scott M, Torchia M, Yang S-L, Serban H, DeArmond SJ, Ebeling C, Prusiner SB, Carlson GA (1991) Paradoxical shortening of scrapie incubation times by expression of prion protein transgenes derived from long incubation period mice. Neuron 7: 59–68

Westaway D, Cooper C, Turner S, Da Costa M, Carlson GA, Prusiner SB (1994a) Structure and polymorphism of the mouse prion protein gene. Proc Natl Acad Sci USA 91: 6418–6422

Westaway D, DeArmond SJ, Cayetano-Canlas J, Groth D, Foster D, Yang S-L, Torchia M, Carlson GA, Prusiner SB (1994b) Degeneration of skeletal muscle, peripheral nerves, and the central nervous system in transgenic mice overexpressing wild-type prion proteins. Cell 76: 117–129

Wickner RB (1994) Evidence for a prion analog in *S. cerevisiae*: the [URE3] non-Mendelian genetic element as an altered *URE2* protein. Science 264: 566–569

Wilesmith JW, Hoinville LJ, Ryan JBM, Sayers AR (1992a) Bovine spongiform encephalopathy: aspects of the clinical picture and analyses of possible changes 1986–1990. Vet Rec 130: 197–201

Wilesmith JW, Ryan JBM, Hueston WD, Hoinville LJ (1992b) Bovine spongiform encephalopathy: epidemiological features 1985 to 1990. Vet Rec 130: 90–94

Transgenetics and Neuropathology of Prion Diseases

S.J. DEARMOND[1] and S.B. PRUSINER[2]

1 Introduction

Many lines of experimental evidence have converged to show that the etiology and pathogenesis of scrapie in animals and Creutzfeldt-Jakob disease (CJD) and Gerstmann-Sträussler-Scheinker disease (GSS) in humans are causally related to abnormalities of the prion protein (PrP). Furthermore, the infectious prions, which are formed in the sporadic, genetic, and infectious prion diseases, are composed largely, if not exclusively, of an abnormal isoform of PrP (PrP^{Sc}). The ability to mimic GSS in transgenic (Tg) mice expressing a mutant PrP gene and the successful manipulation of all of the characteristics of experimental scrapie in Tg mice by

[1] Departments of Pathology and Neurology, University of California, San Francisco, CA 94143, USA
[2] Departments of Neurology and of Biochemistry and Biophysics, University of California, San Francisco, CA 94143, USA

changing the amino acid sequence of PrP or its level of expression provide persuasive evidence that PrP^{Sc} formation features in both the etiology and pathogenesis of the prion diseases.

Although much progress has been made in deciphering the molecular basis of prion diseases, several important questions remain unanswered. If PrP^{Sc} is the sole functional component of the infectious particle, how does it "code" for different clinical-neuropathological syndromes? A related question is how does a single animal species generate the structural diversity in PrP^{Sc} necessary to form multiple prion "strains"? What is the mechanism of selective vulnerability of neurons as a function of prion strain given that each prion isolate causes a different clinical-pathological scrapie syndrome?

This review focuses on studies of Tg mice that have served to advance our understanding of the pathogenesis of prion diseases. Tg mice expressing foreign and mutant PrP genes have provided a wealth of information about prion biology as well as the diseases caused by prions.

2 PrP^{Sc} is Derived from PrP^{C}

From studies of cell lines which remain stably infected with scrapie after multiple passages in culture, it has been found that PrP^{Sc} is derived from the constitutively expressed normal PrP isoform, PrP^{C}. The majority of PrP^{C} becomes attached to the outer surface of these cells by a glycolipid anchor (GPI anchor) after it passes through the Golgi. As with other GPI anchored proteins, mouse (Mo) PrP^{C} reenters the cell through caveolae (TARABOULOS et al. 1995; YING et al. 1992). Some investigators argue that chicken PrP, which is approx. 30% homologous with mammalian PrP, is internalized into mouse neuroblastoma cells through clathrin-coated pits, and that a small proportion of chicken PrP^{C} which is endocytosed is degraded while the majority is recycled to the cell surface along with other proteins and lipids of the endosome (SHYNG et al. 1994). In pulse-chase experiments PrP^{C} in uninfected cells is rapidly labeled by a radioactive amino acid tracer and appears to be degraded in about 6 h (BORCHELT et al. 1990; CAUGHEY et al. 1989). In scrapie-infected cells labeling of PrP^{Sc} is delayed by about 1 h after the pulse and increases to a maximum during the time period when the PrP^{C} pool loses the tracer (BORCHELT et al. 1990). Over the next 24 h there is little or no loss of tracer from the PrP^{Sc} pool. These results argue that PrP^{Sc} is derived from PrP^{C}, but, unlike PrP^{C}, PrP^{Sc} is not degraded and accumulates in the cell. Immunoelectron microscopy studies indicate that PrP^{Sc} accumulates in structures believed to be secondary lysosomes containing "myelin" figures (e.g., layers of phospholipid-rich membranes) (MCKINLEY et al. 1990).

In vivo the same process appears to occur most probably in CNS neurons since they express the highest levels of PrP mRNA (DEARMOND et al. 1992; KRETZSCHMAR et al. 1986a). Following intracerebral inoculation of an animal with prions, PrP^{Sc} begins to accumulate exponentially in the brain. Of interest, PrP^{Sc} concentration is proportional to the scrapie infectivity titer (JENDROSKA et al. 1991), consistent with the hypothesis that it is a component of the prion. Furthermore, the characteristic neuropathological features of prion diseases, which are spongiform degeneration of neurons, nerve cell loss and reactive astrocytic gliosis, colocalize with the sites of PrP^{Sc} accumulation, consistent with the hypothesis that PrP^{Sc} accumulation causes neuronal dysfunction and neuropathology (DEARMOND and PRUSINER 1993).

3 Similarities and Differences Between PrP^{Sc} and PrP^{C}

Multiple attempts over several years to find a chemical difference which could distinguish PrP^{Sc} from PrP^{C} have failed. They have identical molecular weights of 33–35 kDa. They have the same amino acid sequence, similar N-linked carbohydrate side chains, and GPI anchors (STAHL et al. 1993). In spite of these similarities PrP^{Sc}, and similarly PrP^{CJD} which accumulates in human prion diseases, can be distinguished from PrP^{C} in several ways. PrP^{Sc} is associated with scrapie infectivity whereas PrP^{C} is not. PrP^{C} is completely digested by limited exposure to proteinase K whereas PrP^{Sc} and PrP^{CJD} are partially digested to a 27- to 30-kDa protein designated PrP 27–30 which is equally infectious as PrP^{Sc}. PrP 27–30 derived from PrP^{Sc} or PrP^{CJD} aggregates into rodlike structures when exposed to detergents, which are morphologically indistinguishable from many purified amyloids both ultrastructurally and tinctorially.

The only known posttranslational modification of PrP^{C} which occurs during its transformation to PrP^{Sc} is the acquisition of β-sheet conformation (PRUSINER et al. 1983). PrP^{Sc} is 43% β-sheet and 30% α-helix whereas PrP^{C} is 3% β-sheet and 42% α-helix (PAN et al. 1993). PrP 27–30 has an even higher β-sheet content (54%) and lower α-helix content (21%) than PrP^{Sc}, which presumably accounts for its propensity to form into amyloid rods (CAUGHEY et al. 1991; GASSET et al. 1993). Similarly, limited proteolysis PrP^{Sc} and PrP^{CJD}, both of which possess high β-sheet content, presumably account, at least in part, for cerebral amyloidogenesis in prion diseases.

4 Inherited Prion Diseases in Humans

With the discovery of the prion protein (BOLTON et al. 1982; PRUSINER et al. 1982) and subsequent recovery of human PrP cDNAs (KRETZSCHMAR et al. 1986b), molecular genetic studies of familial prion disease became possible (Fig. 1). The human PrP gene designated PRNP has been found to contain mutations in

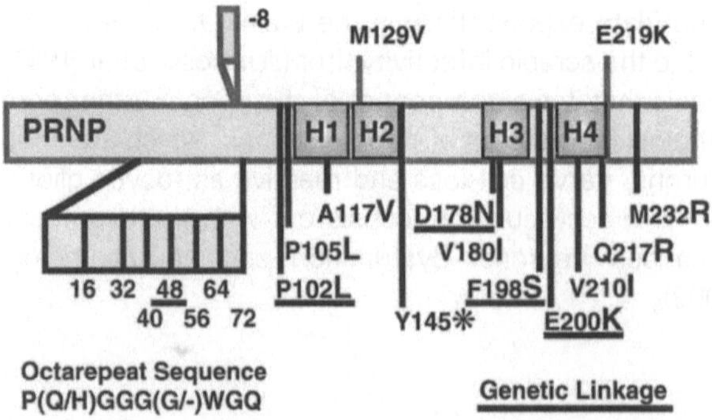

Fig. 1. Mutations that segregate with inherited prion diseases (*below*) and wild-type polymorphisms (*above*) of the open reading frame of the human prion protein gene (*PRNP*). *Boxes*, locations of the four putative α-helical regions in the ORF (*H1–H4*; HUANG et al. 1994). The wild-type *PRNP* gene ORF contains five octarepeats from codon 51 to 91 (KRETZSCHMAR et al. 1986b). Deletion of a single octarepeat at codon 81 or 82 is not associated with prion disease (LAPLANCHE et al. 1990; PALMER et al. 1993; PUCKETT et al. 1991; VNENCAK-JONES and PHILLIPS 1992). The normal distribution of the common polymorphism coded at codon 129 in the United States, United Kingdom, and France is: M/V = 51%, M/M = 38%, and V/V = 11% (BROWN et al. 1992; COLLINGE et al. 1991b). The Met allotype at codon 129 appears to be more prevalent among normal Japanese (M:V = 0.958:0.043) than North American and European populations (M:V = 0.625:0.375; MIYAZONO et al. 1992). Homozygosity for Met or Val at codon 129 appears to increase susceptibility to sporadic CJD (COLLINGE et al. 1991b) and to iatrogenic CJD (BROWN et al. 1992; COLLINGE et al. 1991a). *Small rectangles below ORF*, octarepeat inserts of 16, 32, 40, 48, 56, 64, and 72 amino acids at codon 67, 75, or 83 are shown by these inserts segregate with familial CJD (GOLDFARB et al. 1991a; OWEN et al. 1989, 1990). Point mutations which segregate with prion diseases are designated by the wild-type amino acid preceding the codon number followed by the mutant residue. *Underlining*, mutations for which significant genetic linkage has been established because sufficient family members were examined. Mutations at codons 102, 105, 117 have been linked to different GSS pedigrees (DOH-URA et al. 1989; GOLDFARB et al. 1990a; GOLDGABER et al. 1989; HSIAO et al. 1989a,b; KITAMOTO et al. 1993b). Point mutations at codon 198 and 217 have been found in families with GSS and neurofibrillary tangles (DLOUHY et al. 1992; HSIAO et al. 1992; NOCHLIN et al. 1989). GSS has been linked to a point mutation at codon 145 which results in a stop codon (*asterisk*; KITAMOTO et al. 1993a). Familial CJD has been linked to mutations at codons 178, 180, 200, 210, and 232 (GABIZON et al. 1991; GOLDFARB et al. 1990b, 1991b, 1992a; HSIAO et al. 1991; KITAMOTO et al. 1993b; RIPOLL et al. 1993). Pedigrees with FFI have the same D178N mutation as 178-familial CJD families; however, the mutated allele in FFI also codes for a Met at codon 129 whereas the mutated allele in 178-familial CJD has a Val coded at 129 (GOLDFARB et al. 1992b)

families with inherited prion diseases. Genetic linkage studies have demonstrated a causal relationship between five different PRNP mutations and CNS degeneration (DLOUHY et al. 1992; GABIZON et al. 1993; HSIAO et al. 1989a; PETERSEN et al. 1992; POULTER et al. 1992).

The nomenclature used to designate each of these disorders is still evolving. Currently the tendency among clinical investigators is to group the mutations into two categories: those associated with clinical and neuropathological features which more closely resemble CJD and those which more closely resemble GSS. CJD are recognized by subacute progression of dementia and motor signs, myoclonus, periodic triphasic discharges on EEG, and widespread spongiform degeneration in the cerebral gray matter, most often without amyloid plaque formation. GSS are characterized by slowly progressive cerebellar signs, variable spongiform degeneration of gray matter, and, by definition, widespread deposition of PrP immunopositive amyloid plaques. The amyloid deposition in GSS consists of numerous primitive and mature PrP plaques in both the cerebellum and cerebral hemispheres which can be as large as 100–200 µm (Fig. 2). Three mutations associated with GSS occur in the region of the first putative α-helical domain of PrP (GASSET et al. 1992; HUANG et al. 1994) at codons 102, 105, and 117 (Fig. 1). A fourth occurs at codon 145 at the C-terminal end of the second helical domain (KITAMOTO et al. 1993a). This mutation results in a stop codon and the synthesis of truncated PrP. Two other mutations associated with widely deposited PrP amyloid occur in the region of the third and fourth helical domains at codons 198 and 217 (DLOUHY et al. 1992; HSIAO et al. 1992). In both of these disorders PrP amyloid plaque formation in the cerebral cortex is associated with neurofibrillary degeneration of neurons of the type seen with aging and Alzheimer's disease. The codon 198 mutation is linked to a large pedigree in Indiana and is most remarkable because the PrP amyloid deposits in the cerebral cortex form the cores of neuritic plaques of the type characteristic of Alzheimer's disease. CJD has been found in patients with the insertion of variable numbers of octapeptide repeats in N-terminal domain of PrP and mutations at codons 178, 180, 200, 210, and 232 in the region of the third and fourth helical domains. The mutation at codon 178 is of particular interest because it results in an Asn for Asp substitution and is linked to both FFI and a typical CJD syndrome (GOLDFARB et al. 1992a; MEDORI et al. 1992a,b). Subsequently it was discovered that the PRNP allele carrying the 178 mutation in FFI also codes for a Met polymorphism at codon 129 while the mutated allele in familial CJD (D178N) codes for Val at codon 129 (GOLDFARB et al. 1992b). These findings appear to demonstrate how a single amino acid difference in a mutated PrP molecule determines which population of neurons are vulnerable and the resulting clinical features. In FFI (D178N, M129) the neuropathology is confined largely to the mediodorsal and anterior ventral nuclei of the thalamus whereas in familial CJD (D178N, V129) the neuropathology is widespread in the cerebral cortex and subcortical nuclei. An additional distinction is that seven of ten familial CJD (D178N, V129) cases but none of three FFI cases have transmitted to nonhuman primates (BROWN et al. 1994).

Alternatively, some investigators have suggested that these disorders all be designated as inherited prion diseases followed by the mutations given in parentheses (PRUSINER et al. 1994). This approach seems most practical since substantial numbers of patients carrying a particular mutation within a single pedigree have shown different clinical presentations.

5 Tg(MoPrP-P101L) Mice Provide a Model for Human GSS(P102L)

Tg mouse lines expressing a mutant mouse PrP which carry the codon 102 mutation linked to GSS in humans, designated Tg(MoPrP-P101L)H mice, were

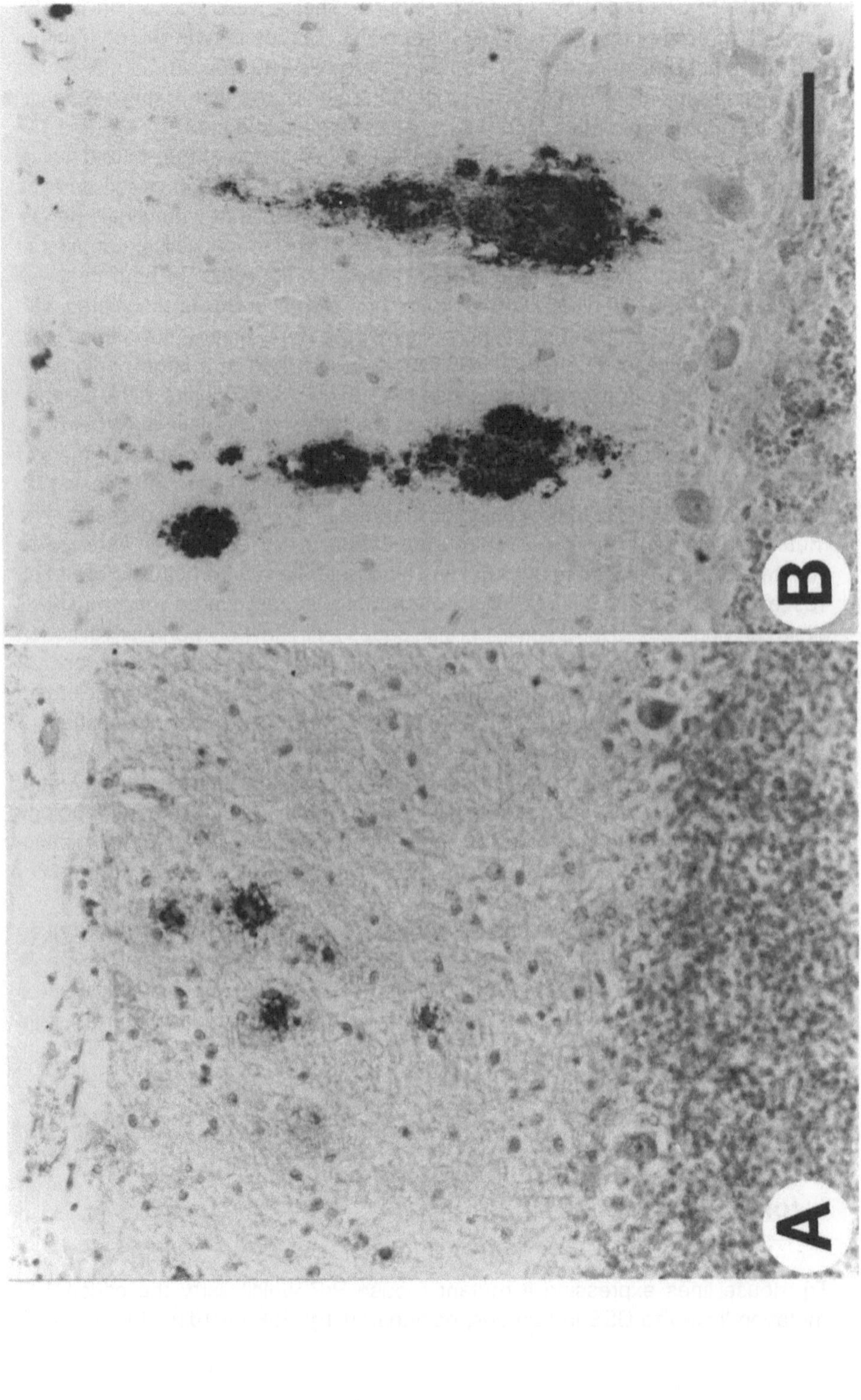

B
A

constructed (HSIAO et al. 1990). Codon 101 in mice is analogous to codon 102 in humans. Two Tg mouse lines expressing high (H) levels of the transgene product developed a spontaneous neurodegenerative disorder with spongiform degeneration of gray matter and formation of PrP immunopositive amyloid plaques in multiple brain regions similar to those found in human GSS (Fig. 3). Thus, the characteristic neuropathological features of GSS were duplicated in these mice. A third line of Tg(MoPrP-P101L) mice expressing low levels of mutant PrP do not develop spontaneous CNS degeneration; these mice are designated Tg196 mice (HSIAO et al. 1994).

The other characteristic feature of human GSS as well as familial CJD is transmissibility, although the rate of successful transmission is different for each PRNP gene mutation. In the NIH series (BROWN et al. 1994) the rate of transmission to primates was highest (85%) for CJD(E200K), 70% for CJD(D178N, V129), relatively unsuccessful (38%) for GSS(P102L), and 0% for three cases of FFI(D178N, M129). Extracts from a dead Japanese patient with GSS(Y145*) (KITAMOTO et al. 1993a) have not transmitted to rodents (T. KITAMOTO, personal communication).

Transmission of brain extracts from Tg(MoPrP-P101L)H mice to Syrian hamsters and Tg196 mice (Fig. 3) expressing low levels of the transgene product has been achieved (HSIAO et al. 1994). Twelve of 164 Syrian hamsters (about 7%) and 73 of 152 Tg196 mice (about 40%) developed a scrapielike neurodegenerative disease after inoculation. These rates of transmission, although relatively low, are significant, and different than the rate of transmission with normal brain homogenates from healthy CD-1 mice or non-Tg littermates. In contrast, only 2 of 97 Tg196 became ill following inoculation with normal homogenates and only after 500 days. Brain homogenates from clinically ill Tg(MoGSS-P101L)H mice did not transmit to CD-1 mice (Table 1). Transmissions to Tg196 mice are of particular interest because these Tg mice also express the MoPrP-P101L transgene, but at much lower copy numbers than the Tg mice which developed diseases spontaneously and do not develop a neurodegenerative disease spontaneously. One explanation for the relatively high rate of transmission to Tg196 from clinically ill Tg(MoPrP-P101L)H mice is that Tg196 expresses the same mutated PrP transgene product.

6 Prnp$^{0/0}$ Mice Neither Develop Scrapie nor Propagate Prions

Mice in which the PrP gene has been ablated have normal life spans without any structural or behavioral abnormalities (BÜELER et al. 1992). After inoculation with

◄

Fig. 2A,B. PrP immunopositive plaques in the cerebellar cortex of two forms of GSS. **A** GSS(P102L). **B** GSS(A117V). Peroxidase immunohistochemistry was facilitated by treating the tissue section by the hydrolytic autoclaving technique (MURAMOTO et al. 1992). *Bar* (**B**), 100 µm. Hematoxylin counterstain

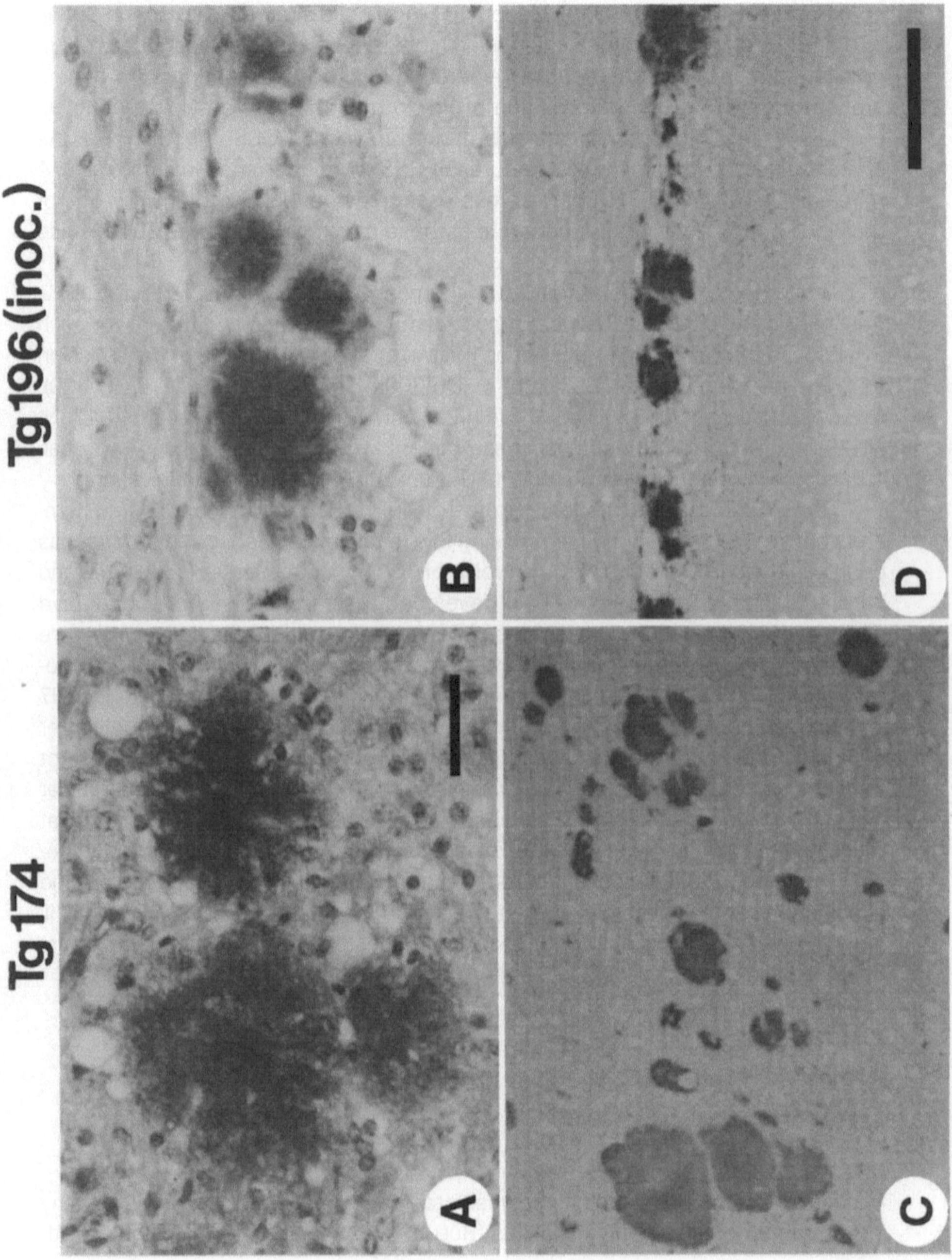
Tg174
Tg196 (inoc.)
A
B
C
D

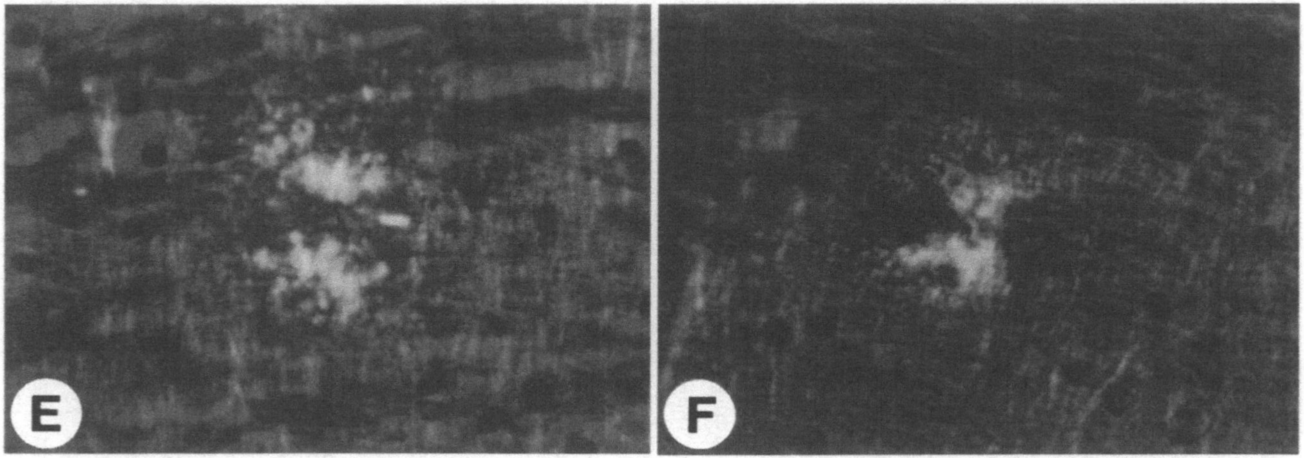

Fig. 3. GSS-type PrP amyloid plaques in the brains of Tg174 (**A**, **C**, **E**) mice which develop a spontaneous neurodegenerative disease and in Tg196 mice (**B**, **D**, **F**) inoculated with a brain homogenate from Tg174. Similarities shared by the plaques were strong histochemical staining by the periodic acid/Schiff method (**A**, **B**), immunoreactivity with PrP-specific antibodies (**C**, **D**), and green-gold birefringence when stained with Congo red and viewed with polarized light (**E**, **F**). The main differences were that Tg174 plaques were primarily in the caudate nucleus and were as large as 90 μm, whereas those in inoculated Tg196 mice were primarily in the subcallosal region (**D**) over the hippocampus and smaller. Peroxidase immunohistochemistry with anti-PrP antibodies (R073) stained amyloid plaques in the caudate nucleus (**C**) and subcallosal plaques (**D**). **B**, **D**, **F** Tg196 mice were inoculated with homogenates prepared from the brains of clinically ill Tg174 mice. **A**, **B**, **E**, **F** *Bar*, 50 μm. **C**, **D** *Bar*, 100 μm

Table 1. Transmission of neurodegeneration to animals by inoculation of brain homogenates from spontaneously ill Tg(MoGSS-P101L) mice (number with neurological disease out of total inoculated) (from HSIAO et al. 1994)

Source of inoculum	CD-1 mice	Syrian hamsters	Tg196 mice
Symptomatic Tg(MoGSS-P101L) mice	0/160	9/148	62/138
Presymptomatic Tg(MoGSS-P101L) mice	0/30	0/24	0/26
Normal non-Tg mice	0/150	0/120	0/88[a]

Mice were inoculated intracerebrally with 30 μl and hamsters with 50 μl 10% brain homogenate.
[a] One animal with scrapielike syndrome 476 days after inoculation.

prions they neither develop clinical nor neuropathological features and do not propagate prions. However, when a Syrian hamster (SHa) PrP transgene was introduced into $Prnp^{0/0}$ mice by crossing them with Tg(SHaPrP) mice, the Tg(SHaPrP)/$Prnp^{0/0}$ mice became susceptible to SHa prions but not to Mo prions (BÜELER et al. 1993; PRUSINER et al. 1993). These findings indicate that the absence of PrP^{C} expression is not the cause of neuronal dysfunction in scrapie and support the view that accumulation of PrP^{Sc} is responsible for CNS degeneration. Furthermore, these data indicate that prion propagation requires PrP^{C}. Finally, the transmission of scrapie with SHa prions containing $SHaPrP^{Sc}$ but not with prions containing $MoPrP^{Sc}$ to mice which express $SHaPrP^{C}$ but not $MoPrP^{C}$ indicates that propagation of prions requires amino acid homology between the PrP^{Sc} of the prion and PrP^{C} of the host.

7 Prion Propagation and the Host Barrier to Scrapie Infection Depends on PrP Amino Acid Sequence

Two prion isolates to which there are well-defined host barriers are the Mo(RML) isolate which was passaged in mice and the SHa(Sc237) isolate which was passaged in Syrian hamsters (PRUSINER et al. 1990; SCOTT et al. 1989). When Syrian hamsters were inoculated in the thalamus with SHa(Sc237) prions, clinical signs of scrapie became detectable in approx. 75 days, whereas no Syrian hamsters became clinically ill in over 500 days with Mo(RML) prions. Conversely, signs of scrapie were detectable in mice about 140 days after intrathalamic inoculation with Mo(RML) prions, but they did not develop signs of disease after more than 500 days when inoculated with SHa(Sc237) prions. That this barrier is due to amino acid differences between PrP^{Sc} in the prion and PrP^{C} of the host has been demonstrated in Tg mice expressing a variety of PrP transgenes.

The specificity of the interaction between PrP^{Sc} of the infecting prion and PrP^{C} of the host was first demonstrated in Tg(SHaPrP) mice which express both $SHaPrP^{C}$ and $MoPrP^{C}$ (PRUSINER et al. 1990; SCOTT et al. 1989). When these animals were inoculated with SHa(Sc237) prions, only $SHaPrP^{Sc}$ was formed, based on the characteristics of the neuropathology, the presence of $SHaPrP^{Sc}$ in the neuropil and in amyloid plaques, and the behavior of the newly formed prions which had characteristics of Sc237 prions (e.g., they infected Syrian hamsters but not mice). In contrast, Mo(RML) prions only interacted with $MoPrP^{C}$ since neuropathology characteristic of RML scrapie developed which is defined by a different distribution of gray matter vacuolation, vacuolation of the white matter, and the absence of amyloid plaques. Thus, Sc237 prions selectively interacted with $SHaPrP^{C}$ in these Tg(SHaPrP) mice, and RML prions selectively interacted with $MoPrP^{C}$.

These observations have greatly increased our understanding of the PrP^{C}-PrP^{Sc} interactions which lead to formation of nascent PrP^{Sc} (Fig. 4). The necessity for amino acid homology between PrP^{C} of the host and PrP^{Sc} within the prion is well established. Considerable data argue that a heterodimer consisting of PrP^{C} and PrP^{Sc} is an intermediate in the formation of nascent prions.

8 Selective Vulnerability of Neurons as a Function of Prion Isolate and Sites of PrP^{Sc} Accumulation

Neurohistopathological examination of the brains of Tg mice has contributed to understanding the role of PrP^{Sc} in the pathogenesis of prion diseases and the origin of clinical-neuropathological diversity. As a consequence, new insights into the mechanism of differential selective vulnerability of neurons as a function of prion isolates have emerged. Two of the characteristics which differentiate prion isolates, the distribution and intensity of spongiform degeneration and formation of amyloid plaques, have been found to be determined largely, if not entirely, by PrP^{Sc}. Amyloid plaques in prion disorders contain PrP (DEARMOND et al. 1985). Localization of PrP^{Sc} in the brain by immunohistochemistry and histoblot analysis has revealed a precise topographical correlation between the sites of PrP^{Sc} accumulation and pathological changes characterized by spongiform degeneration and reactive astrocytic gliosis. This relationship has been found with the Sc237 and 139H isolates in Syrian hamsters (DEARMOND et al. 1987; HECKER et al.

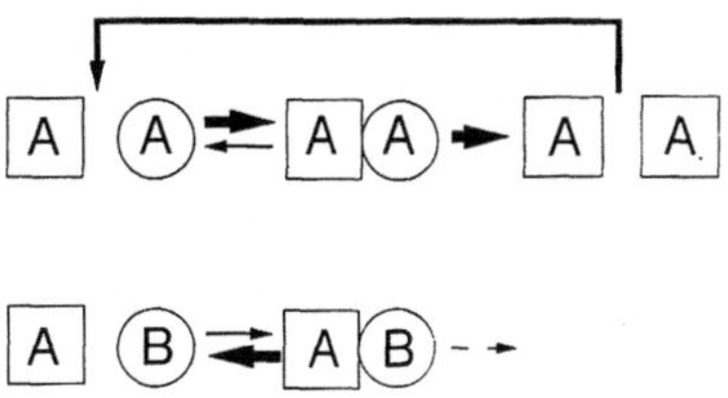

Fig. 4. Schematic diagram illustrating the dimer hypothesis of prion propagation and the host species barrier to scrapie infection. *Squares*, PrP^{Sc}; *circles*, PrP^{C}. When there is sufficient amino acid homology between PrP^{Sc} (*A*) and PrP^{C} (*A*), dimer formation is favored and is followed by transformation of PrP^{C} (*A*) into a duplicate copy of PrP^{Sc} (*A*). In the absence of sufficient amino acid homology, PrP^{Sc} (*A*) tends not to bind to PrP^{C} (*B*) with the result that PrP^{C} (*B*) is not transformed into nascent PrP^{Sc} (*A*)

1992) and with the Sc237, 139H and isolates in Tg(SHaPrP) mice (DEARMOND et al. 1993; HECKER et al. 1992). Similarly, the same relationship has been found with the Sc237 and RML isolates in Tg(MH2M) mice expressing a chimeric PrP containing both Mo and SHaPrP sequences (SCOTT et al. 1993) and with Hu CJD prions inoculated into Tg(MHu2M) mice expressing a chimeric PrP containing both Mo and HuPrP sequences (TELLING et al. 1994). In addition to this spatial relationship, there is also a temporal correlation between the accumulation of PrP^{Sc} in a brain region and the subsequent development of neuropathology (JENDROSKA et al. 1991).

Because PrP^{Sc} accumulation appears to cause spongiform degeneration, and because the distribution of spongiform degeneration is characteristic for each prion isolate, it follows that the distribution of PrP^{Sc} must also be characteristic of each prion isolate. This relationship was established when the highly sensitive histoblot technique which allows PrP^{C} to be completely eliminated from the tissue section by exposure of the histoblot to proteinase K and subsequent exposure to guanidinium which denatures PrP^{Sc} and markedly increases its affinity for PrP antibodies was developed (TARABOULOS et al. 1992). Earlier studies using PrP immunohistochemical techniques on aldehyde-fixed tissues were interpreted similarly but the immunostaining procedure used failed to eliminate PrP^{C} (BRUCE et al. 1989).

The histoblot technique has clearly revealed that the neuroanatomic location of PrP^{Sc} accumulation is unique for each scrapie prion isolate and not a function of scrapie incubation time when compared in animals with the same genotype (DEARMOND and PRUSINER 1993; DEARMOND et al. 1993; HECKER et al. 1992). In comparing the kinetics of PrP^{Sc} accumulation in the brain of Syrian hamsters inoculated with either Sc237 or 139H prions, we found that its distribution at the time clinical signs became apparent was markedly different (Fig. 5). Specifically, PrP^{Sc} was localized to fewer brain regions with Sc237 than with 139H; however, because scrapie incubation time was significantly longer with 139H, 170 days versus 70 days for Sc237, we could not rule out the possibility that the longer incubation time with 139H permitted the disease to spread nonspecifically to more brain regions rather than showing selective targeting. The influence of scrapie incubation time was eliminated in studies with Tg(SHaPrP)7 mice that express high levels of SHaPrP mRNA and have scrapie incubation times of about 50 days with both the Sc237 and 139H isolates (HECKER et al. 1992). After Sc237 prions were inoculated unilaterally into the thalamus, PrP^{Sc} accumulation was largely confined to selected nuclei in the thalamus, septum, and brainstem while it was widely distributed throughout the CNS with 139H (Fig. 5). With the Me7H isolate, PrP^{Sc} accumulation was even more restricted being confined to the paraventricular nucleus of the thalamus, the habenula, hypothalamus, zona incerta, nucleus accumbens septi, and periaqueductal gray of the midbrain. These findings were all the more remarkable because scrapie incubation times with Me7H in Tg(SHaPrP)7 mice were about 185 days, and therefore the highly restricted distribution of PrP^{Sc} could not be attributed to insufficient time for prions to spread within the CNS.

9 Artificial Prions Formed from Chimeric SHa/MoPrP^C

The homotypic interaction of PrP^{Sc} with PrP^C was also demonstrated in Tg mice expressing chimeric PrP^C containing amino acid sequences from both SHa and MoPrP (Scott et al. 1993). Two chimeric constructs were made on the MoPrP background: one containing two amino acid substitutions from SHa at codons 108 and 111 designated MHM2 PrP and the other containing three additional substitutions at codons 138, 154, and 169 designated MH2M. Three Tg mouse lines expressing the former construct, Tg(MHM2) mice, were resistant to SHa(Sc237) derived from Syrian hamsters similar to non-Tg mice; however, all Tg mice expressing the transgene with five SHa amino acid substitutions, Tg(MH2M) mice, became clinically ill with SHa(Sc237) prions. This argues that the homophilic interaction between PrP^{Sc} in the prion and PrP^C in the host leading to a breach of the species barrier and clinical disease is related to the amino acid sequence of both molecules. When the resulting prions, designated MH2M(Sc237), were passaged back into Syrian hamsters, they developed scrapie characterized by distribution of PrP^{Sc} similar to, although not identical with, that caused by SHa(Sc237). Mouse-derived RML prions, Mo(RML), also produced scrapie in the Tg(MH2M) mice. Although Mo(RML) prions do not infect Syrian hamsters, inoculation of brain homogenates infected with Mo(RML) prions from Tg(MH2M) mice into Syrian hamsters did produce scrapie indicating that MH2M(RML) prions had been formed. Furthermore, the pattern of PrP^{Sc} accumulation in Syrian hamster brain was unique, suggesting that the artificial prion created by the chimeric transgene was a new scrapie prion isolate. Particularly unique aspects of the PrP^{Sc} pattern were its dense accumulation along the hippocampal fissure and subpially at the periphery of the brain. Both the MH2M(RML) and MH2M(Sc237) prions also transmitted to CD-1 mice. The successful transmission of MH2M(Sc237) prions to mice indicates that the amino acid sequence of this chimeric prion was able to overcome the natural mouse species barrier to SHa(Sc237) prions containing SHaPrP^{Sc}. These findings also argue that the amino acid sequence homology in the region of the first putative α-helical domain of the PrP molecule may be particularly important in PrP^{Sc}/PrP^C dimerization.

10 Amino Acid Sequence Homology Insufficient for Crossing the Prion Species Barrier Between Humans and Mice

That there is another factor or set of factors which determine whether a prion isolate initiates the transformation of PrP^C to nascent PrP^{Sc} became evident when we found that transmission of HuCJD prions to Tg(HuPrP) mice, which express HuPrP^C, occurred in only about 10% of the inoculated mice and was similar to our

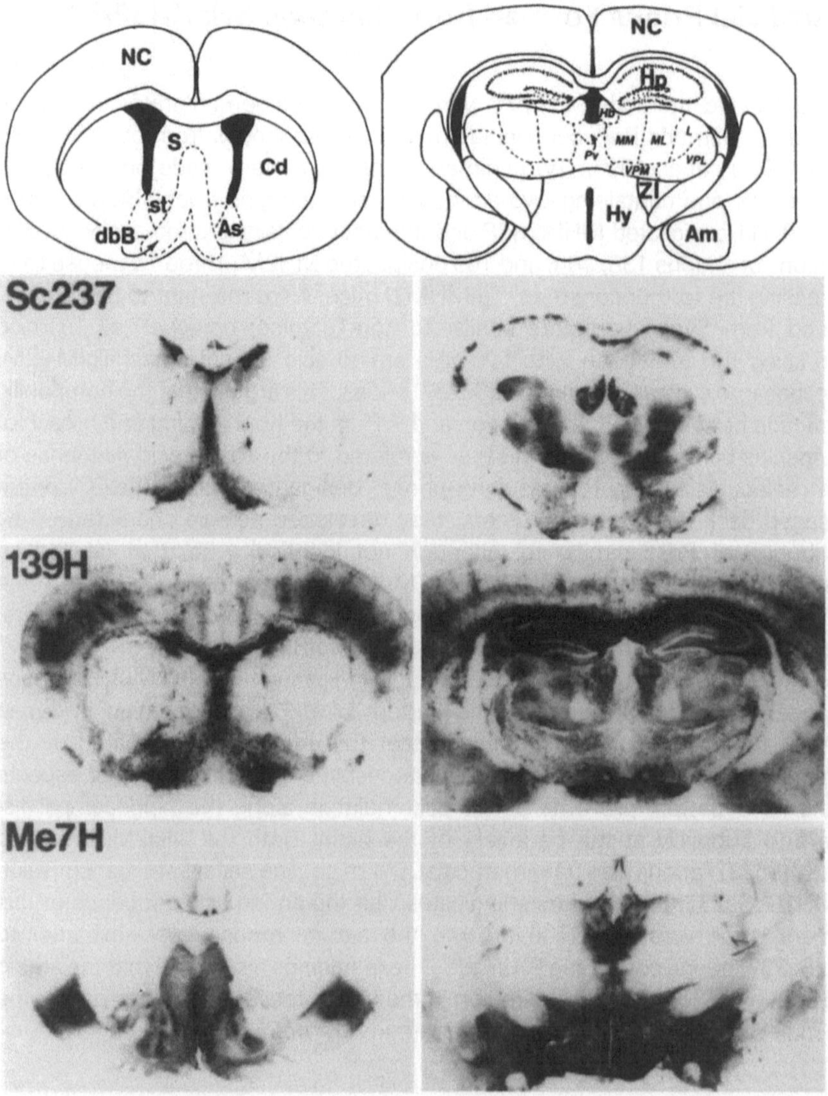

Fig. 5. The distribution of PrP^{Sc} in the brains of Tg(SHaPrP) mice is significantly different for the Sc237, 139H, and Me7H prion isolates; however, there are brain regions which accumulate PrP^{Sc} regardless of the prion isolate such as the diagonal band of Broca (*dbB*), the hypothalamus (*Hy*), and various brainstem nuclei

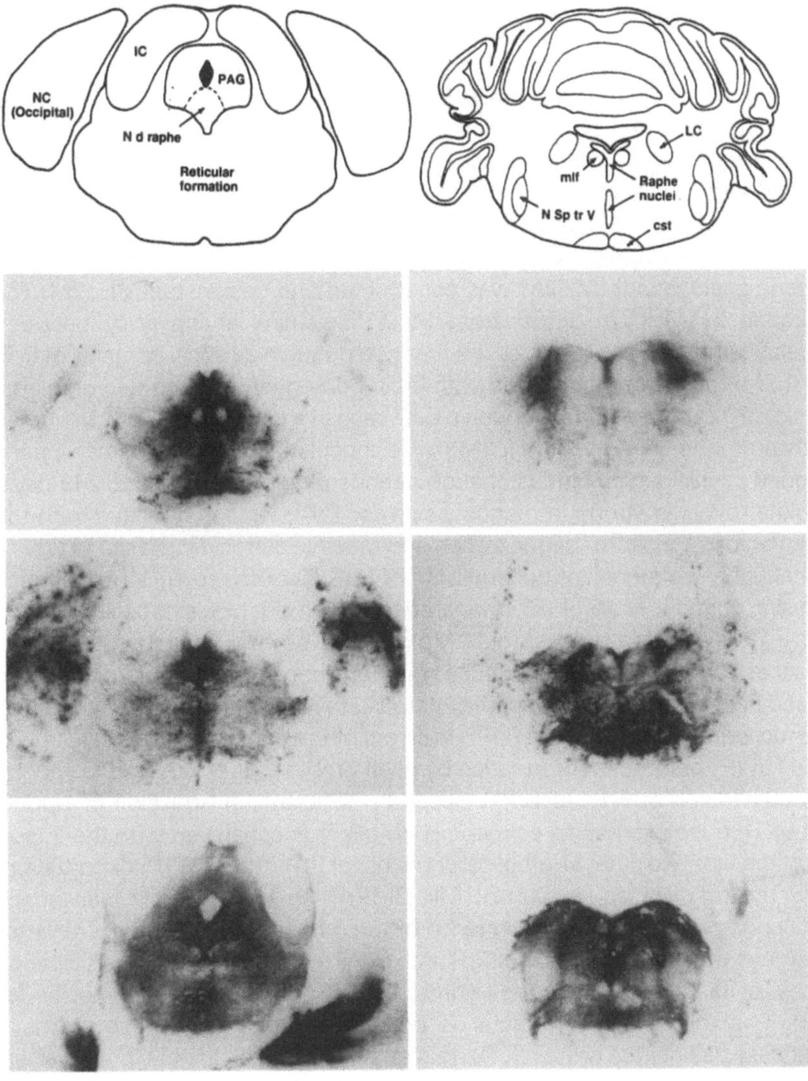
IC
PAG
NC
(Occipital)
N d raphe
Reticular
formation
LC
mlf
Raphe
nuclei
N Sp tr V
cst

Fig. 5 (*Contd.*)

findings with non-Tg mice (TELLING et al. 1994). Thus when two Tg(HuPrP) mouse lines were inoculated intrathalamically with brain extracts from 18 patients who had died of sporadic CJD, iatrogenic CJD, familial CJD, or GSS over a 2.5-year period, only 14 of 169 Tg(HuPrP) mice became clinically ill (8.3%), which is similar to the 6 of 58 control non-Tg mice which had become clinically ill (10.3%). Incubation times were extremely long, ranging from 590 to 840 days, for both Tg(HuPrP) and non-Tg mice.

Because of the successful transmission of scrapie with both SHa(Sc237) and Mo(RML) prions in Tg mice expressing chimeric SHa/MoPrP^{C}, a chimeric Hu/Mo transgene analogous to MH2M was constructed and designated MHu2M. The PrP^{C} coded by MHu2M differs from MoPrP^{C} by nine amino acids between residues 96 and 167, the region of the first two putative α-helical domains of PrP. Tg(MHu2M) mice were inoculated with brain homogenates from two unrelated sporadic CJD cases and one iatrogenic CJD case in a patient treated with human growth hormone. All 24 Tg(MHu2M) mice inoculated with the human prions developed clinical signs with incubation periods ranging from 202 to 249 days. Histoblots revealed strong immunostaining for PrP^{Sc} which correlated with the distribution of spongiform degeneration and reactive astrocytic gliosis.

The higher frequency of transmission of HuCJD prions to Tg(MHu2M) mice expressing chimeric Hu/MoPrP^{C} compared to Tg(HuPrP) mice expressing HuPrP^{C} or non-Tg mice which express only MoPrP^{C} has several implications. First, it argues that 100% amino acid sequence homology between PrP^{Sc} (or PrP^{CJD}) of the prion and PrP^{C} of the host is not sufficient to sustain the self-propagating process of PrP^{C} conversion into nascent PrP^{Sc}. In terms of the dimer hypothesis, it argues that PrP^{Sc}/PrP^{C} heterodimer formation by itself probably does not lead to spontaneous conversion of PrP^{C} to PrP^{Sc}. Secondly, efficient transmission of HuCJD prions to Tg mice expressing chimeric Hu/MoPrP^{C} is consistent with the importance of the first two putative α-helical domains for PrP^{Sc}(PrP^{CJD})/PrP^{C} dimerization. Thirdly, the efficient conversion of MHu2MPrP^{C} to MHu2MPrP^{Sc} initiated by HuCJD prions argues for the importance of the mouse PrP sequence. The latter point argues that the conversion of PrP^{C} into PrP^{Sc} requires the active participation of a cellular factor that recognizes some residues of PrP^{C} that are specific for MoPrP. This mouse specific factor is likely to be a protein which we have provisionally designated protein X (TELLING et al. 1994). Presumably, protein X is a chaperone-like molecule that forms a ternary complex with PrP^{C} and PrP^{Sc} and then catalyzes the conversion of PrP^{C} into PrP^{Sc}. We assume that there is a sizeable energy barrier which separates the α-helical conformational state of PrP^{C} from the β-sheet state of PrP^{Sc} (COHEN et al. 1994). Whether specific prion "strain" information is also transferred to nascent prions during the transformation of PrP^{C} into PrP^{Sc} remains to be established. The most likely explanation for the successful transmission of SHa(Sc237) prions to Tg(SHaPrP) mice expressing SHaPrP^{C} and the failure of transmission of HuCJD prions to Tg(HuPrP) mice expressing HuPrP^{C} is the relative degree of homology of SHaPrP and HuPrP with MoPrP: SHaPrP differs from MoPrP at 16 amino acids whereas HuPrP differs at 28 (WESTAWAY et al. 1987).

11 Transgenic Mice and Scrapie Incubation Times

Depending upon the particular species or strain of animal, prion isolates may be distinguished by their scrapie incubation times. That PrP controls the length of the incubation time is now well established, which in turn argues that distinct prion isolates produce different incubation times through PrP.

Scrapie incubation times in Tg(SHaPrP) mice following inoculation with SHa Sc237 prions were found to be inversely proportional to the level of SHaPrP mRNA expression and of SHaPrP^{C} in four Tg(SHaPrP) mouse lines (PRUSINER et al. 1990). Thus, the greater the degree of SHaPrP^{C} expression, the shorter the incubation time is.

More than 25 years ago DICKINSON and his colleagues described strains of mice with short and long scrapie incubation times (DICKINSON and MEIKLE 1971; DICKINSON et al. 1968). For example, they found that VM mice had a particularly prolonged incubation time (300 days or more), and that C57BL mice had a relatively short incubation time (180–200 days) with the Me7 scrapie isolate. Scrapie incubation times in the F1 (VM × C57BL) heterozygote mice were intermediate. Three other prion isolates had a significantly different effect on incubation times of which the 22A scrapie isolate was the premiere example. In VM mice scrapie incubation times with the 22A group were similar to the Me7 isolate, about 300 days. However, incubation times were not shorter in C57BL mice; rather they ranged from about 450 to 650 days. Another peculiarity of the 22A group was that incubation times in F1 heterozygote mice were not intermediate between the long and short incubation times of the homozygotes; rather they were longer than the prolonged incubation times in C57BL. This phenomenon was designated "overdominance" (DICKINSON and OUTRAM 1979).

With the discovery of the prion protein and the subsequent recovery of PrP cDNA clones, a genetic linkage study with similar mice was performed. Those studies demonstrated genetic linkage between the PrP gene and scrapie incubation times (CARLSON et al. 1986). DNA sequencing showed that the PrP genes of mice with short (PrP-A) and long (PrP-B) scrapie incubation times differ at codons 108 and 189, producing nonconservative amino acid substitutions (WESTAWAY et al. 1987). These results argued that the PrP gene denoted *Prn-p* in mice and the mouse incubation time designated Prn-i were congruent. Because long Prn-i alleles were thought to encode dominant traits, such as long incubation times, Tg mice harboring PrP transgenes (PrP-B) cloned from a long incubation time mouse, I/LnJ, were constructed. The fertilized eggs used to construct these Tg(PrP-B) mice were from mice with short scrapie incubation times carrying the PrP-A gene. Unexpectedly, the Tg(PrP-B) mice had incubation times shorter than those of the non-Tg parental mice that were homozygous for the PrP-A gene (WESTAWAY et al. 1991).

To resolve what appeared to be paradoxical shortening of scrapie incubation times in Tg(PrP-B) mice, we constructed multiple transgenic and congenic mouse lines with varying numbers of PrP-A and PrP-B alleles. Inoculation of these lines with different prion isolates demonstrated that long Incubation times are not dominant as thought for more than two decades, but rather they are due to a PrP gene dosage effect. With the RML isolate, which is similar to the Me7 isolate used by DICKINSON et al., scrapie incubation times were found to be determined by the number of copies of PrP-A and PrP-B alleles. Because the PrP-A was more efficiently converted into PrP^{Sc} than was PrP-B, incubation times after inoculation of the RML isolate into (C57BL x I/LnJ)F1 mice carrying one copy of each PrP allele were substantially longer than those in the parental C57BL mice with two copies of the PrP-A allele. Similar analyses seem to be equally applicable for other prion isolates such as 22A except that mice inoculated with these isolates preferentially convert PrP-B into PrP^{Sc}, and PrP-A often inhibits PrP^{Sc} formation.

12 Conclusions

An understanding of the molecular mechanisms of both infectious and genetic prion diseases is beginning to emerge from the studies of Tg mice described in this review. The results of these studies argue that mutated PrP as well as PrP^{Sc} (or PrP^{CJD}) plays a preeminent role in both the etiology and pathogenesis of all forms of prion disease. Thus, Tg mice expressing the mutated PrP which mimics the mutation genetically linked to human ataxic GSS(P102L) results in a spontaneous cerebral amyloidosis which reproduces the neuropathological features of the human disorder. Even variability of age of onset and clinical patterns were duplicated when the mutant MoPrP(P101L) was expressed with wt MoPrP. In the absence of wt MoPrP, a more synchronous neurodegenerative disorder has been found indicating that the presence of $wtPrP^{C}$ influences the age of onset (TELLING, in preparation). Even for infectious scrapie, studies with Tg mice have shown that murine prion isolates have differential interactions with MoPrP-A and MoPrP-B alleles which result in different clinical and neuropathological phenotypes (CARLSON et al. 1994). In addition, Tg mice have shown that the host species barrier to a scrapie prion isolate is determined by lack of sufficient amino acid homology between PrP^{C} of the host and PrP^{Sc} of the prion (TELLING et al. 1994). The finding that all clinical and neuropathological aspects of genetic and infectious prion diseases can be changed or eliminated by manipulating the prion protein and the fact that the purified scrapie agent is composed almost exclusively of PrP^{Sc} are persuasive arguments that PrP^{Sc} is likely to be the sole functional component of prions, and that its accumulation in the brain is pathogenic.

The etiology of sporadic CJD is unknown. One possibility is suggested by the molecular genetics of dominantly inherited prion diseases which indicate that many PrP gene point mutations are pathogenic and are associated with spontaneous formation of infectious prions. Therefore, some cases of sporadic CJD may be related to acquired pathogenic mutations in the PrP gene. Such a mutation even in a single cell could trigger a self-propagating formation of pathogenic PrP^{Sc} that spreads throughout the CNS. A second possibility has been suggested by Tg mice which overexpress wtPrP^{C}. Many of these animals spontaneously develop an age-related neuromuscular disorder (WESTAWAY et al. 1994). Focal spongiform degeneration of the hippocampus was also identified. Although no protease resistant SHaPrP was detected by immunoblotting brain homogenates from Tg(SHaPrP)7 mice, these homogenates did transmit scrapie to Syrian hamsters. These findings suggest that prions might form spontaneously in an individual if the rate of degradation or synthesis of PrP^{C} is altered in a particular population of cells. These data also suggest that not all forms of PrP contained within infectious prion particles are protease resistant. In conclusion, a large body of data supports the hypothesis that prions are novel pathogens, and that PrP features in the pathogenesis of the genetic, infectious, and sporadic prion diseases.

References

Bolton DC, McKinley MP, Prusiner SB (1982) Identification of a protein that purifies with the scrapie prion. Science 218: 1309–1311

Borchelt DR, Scott M, Taraboulos A, Stahl N, Prusiner SB (1990) Scrapie and cellular prion proteins differ in their kinetics of synthesis and topology in cultured cells. J Cell Biol 110: 743–752

Brown P, Preece MA, Will RG (1992) "Friendly fire" in medicine: hormones, homografts, and Creutzfeldt-Jakob disease. Lancet 340: 24–27

Brown P, Gibbs CJ Jr, Rodgers-Johnson P, Asher DM, Sulima MP, Bacote A, Goldfarb LG, Gajdusek DC (1994) Human spongiform encephalopathy: the National Institutes of Health series of 300 cases of experimentally transmitted disease. Ann Neurol 35: 513–529

Bruce ME, McBride PA, Farquhar CF (1989) Precise targeting of the pathology of the sialoglycoprotein, PrP, and vacuolar degeneration in mouse scrapie. Neurosci Lett 102: 1–6

Büeler H, Fischer M, Lang Y, Bluethmann H, Lipp H-P, DeArmond SJ, Prusiner SB, Aguet M, Weissmann C (1992) Normal development and behaviour of mice lacking the neuronal cell-surface PrP protein. Nature 356: 577–582

Büeler H, Aguzzi A, Sailer A, Greiner R-A, Autenried P, Aguet M, Weissmann C (1993) Mice devoid of PrP are resistant to scrapie. Cell 73: 1339–1347

Carlson GA, Kingsbury DT, Goodman PA, Coleman S, Marshall ST, DeArmond SJ, Westaway D, Prusiner SB (1986) Linkage of prion protein and scrapie incubation time genes. Cell 46: 503–511

Carlson GA, Ebeling C, Yang S-L, Telling G, Torchia M, Groth D, Westaway D, DeArmond SJ, Prusiner SB (1994) Prion isolate specified allotypic interactions between the cellular and scrapie prion proteins in congenic and transgenic mice. Proc Natl Acad Sci USA 91: 5690–5694

Caughey B, Race RE, Ernst D, Buchmeier MJ, Chesebro B (1989) Prion protein biosynthesis in scrapie-infected and uninfected neuroblastoma cells. J Virol 63: 175–181

Caughey BW, Dong A, Bhat KS, Ernst D, Hayes SF, Caughey WS (1991) Secondary structure analysis of the scrapie-associated protein PrP 27–30 in water by infrared spectroscopy. Biochemistry 30: 7672–7680

Cohen FE, Pan K-M, Huang Z, Baldwin M, Fletterick RJ, Prusiner SB (1994) Structural clues to prion replication. Science 264: 530–531

Collinge J, Palmer M, Dryden A, Campbell T (1991a) Molecular genetics of inherited, sporadic and iatrogenic prion disease. Prion Diseases in Humans and Animals Conference, London

Collinge J, Palmer MS, Dryden AJ (1991b) Genetic predisposition to iatrogenic Creutzfeldt-Jakob disease. Lancet 337: 1441–1442

DeArmond SJ, Prusiner SB (1993) The neurochemistry of prion diseases. J Neurochem 61: 1589–1601

DeArmond SJ, McKinley MP, Barry RA, Braunfeld MB, McColloch JR, Prusiner SB (1985) Identification of prion amyloid filaments in scrapie-infected brain. Cell 41: 221–235

DeArmond SJ, Mobley WC, DeMott DL, Barry RA, Beckstead JH, Prusiner SB (1987) Changes in the localization of brain prion proteins during scrapie infection. Neurology 37: 1271–1280
DeArmond SJ, Jendroska K, Yang S-L, Taraboulos A, Hecker R, Hsiao K, Stowring L, Scott M, Prusiner SB (1992) Scrapie prion protein accumulation correlates with neuropathology and incubation times in hamsters and transgenic mice. In: Prusiner SB, Collinge J, Powell J, Anderton B (eds) Prion diseases of humans and animals. Horwood, London, pp 483–496
DeArmond SJ, Yang S-L, Lee A, Bowler R, Taraboulos A, Groth D, Prusiner SB (1993) Three scrapie prion isolates exhibit different accumulation patterns of the prion protein scrapie isoform. Proc Natl Acad Sci USA 90: 6449–6453
Dickinson AG, Meikle VMH (1971) Host-genotype and agent effects in scrapie incubation: change in allelic interaction with different strains of agent. Mol Gen Genet 112: 73–79
Dickinson AG, Outram GW (1979) The scrapie replication-site hypothesis and its implications for pathogenesis. In: Prusiner SB, Hadlow WJ (eds) Slow transmissible diseases of the nervous system, vol 2. Academic, New York, pp 13–31
Dickinson AG, Meikle VMH, Fraser H (1968) Identification of a gene which controls the incubation period of some strains of scrapie agent in mice. J Comp Pathol 78: 293–299
Dlouhy SR, Hsiao K, Farlow MR, Foroud T, Conneally PM, Johnson P, Prusiner SB, Hodes ME, Ghetti B (1992) Linkage of the Indiana kindred of Gerstmann-Sträussler-Scheinker disease to the prion protein gene. Nat Genet 1: 64–67
Doh-ura K, Tateishi J, Sasaki H, Kitamoto T, Sakaki y (1989) Pro → Leu change at position 102 of prion protein is the most common but not the sole mutation related to Gerstmann-Sträussler syndrome. Biochem Biophys Res Commun 163: 974–979
Gabizon R, Meiner Z, Cass C, Kahana E, Kahana I, Avrahami D, Abramsky O, Scarlato G, Prusiner SB, Hsiao KK (1991) Prion protein gene mutation in Libyan Jews with Creutzfeldt-Jakob disease. Neurology 41: 160
Gabizon R, Rosenmann H, Meiner Z, Kahana I, Kahana E, Shugart Y, Ott J, Prusiner SB (1993) Mutation and polymorphism of the prion protein gene in Libyan Jews with Creutzfeldt-Jakob disease. Am J Hum Genet 33: 828–835
Gasset M, Baldwin MA, Lloyd D, Gabriel J-M, Holtzman DM, Cohen F, Fletterick R, Prusiner SB (1992) Predicted α-helical regions of the prion protein when synthesized as peptides form amyloid. Proc Natl Acad Sci USA 89: 10940–10944
Gasset M, Baldwin MA, Fletterick RJ, Prusiner SB (1993) Perturbation of the secondary structure of the scrapie prion protein under conditions associated with changes in infectivity. Proc Natl Acad Sci USA 90: 1–5
Goldfarb LG, Brown P, Goldgaber D, Asher DM, Rubenstein R, Brown WT, Piccardo P, Kascsak RJ, Boellaard JW, Gajdusek DC (1990a) Creutzfeldt-Jakob disease and kuru patients lack a mutation consistently found in the Gerstmann-Sträussler-Scheinker syndrome. Exp Neurol 108: 247–250
Goldfarb LG, Mitrova E, Brown P, Toh BH, Gajdusek DC (1990b) Mutation in codon 200 of scrapie amyloid protein gene in two clusters of Creutzfeldt-Jakob disease in Slovakia. Lancet 336: 514–515
Goldfarb LG, Brown P, McCombie WR, Goldgaber D, Swergold GD, Wills PR, Cervenakova L, Baron H, Gibbs CJJ, Gajdusek DC (1991a) Transmissible familial Creutzfeldt-Jakob disease associated with five, seven, and eight extra octapeptide coding repeats in the *PRNP* gene. Proc Natl Acad Sci USA 88: 10926–10930
Goldfarb LG, Haltia M, Brown P, Nieto A, Kovanen J, McCombie WR, Trapp S, Gajdusek DC (1991b) New mutation in scrapie amyloid precursor gene (at codon 178) in Finnish Creutzfeldt-Jakob kindred. Lancet 337: 425
Goldfarb LG, Brown P, Haltia M, Cathala F, McCombie WR, Kovanen J, Cervenakova L, Goldin L, Nieto A, Godec MS, Asher DM, Gajdusek DC (1992a) Creutzfeldt-Jakob disease cosegregates with the codon 178^{Asn} *PRNP* mutation in families of European origin. Ann Neurol 31: 274–281
Goldfarb LG, Petersen RB, Tabaton M, Brown P, LeBlanc AC, Montagna P, Cortelli P, Julien J, Vital C, Pendelbury WW, Haltia M, Wills PR, Hauw JJ, McKeever PE, Monari L, Schrank B, Swergold GD, Autilio-Gambetti L, Gajdusek DC, Lugaresi E, Gambetti P (1992b) Fatal familial insomnia and familial Creutzfeldt-Jakob disease: disease phenotype determined by a DNA polymorphism. Science 258: 806–808
Goldgaber D, Goldfarb LG, Brown P, Asher DM, Brown WT, Lin S, Teener JW, Feinstone SM, Rubenstein R, Kascsak RJ, Boellard JW, Gajdusek DC (1989) Mutations in familial Creutzfeldt-Jakob disease and Gerstmann-Sträussler-Scheinker's syndrome. Exp Neurol 106: 204–206
Hecker R, Taraboulos A, Scott M, Pan K-M, Torchia M, Jendroska K, DeArmond SJ, Prusiner SB (1992) Replication of distinct prion isolates is region specific in brains of transgenic mice and hamsters. Genes Dev 6: 1213–1228

Hsiao K, Baker HF, Crow TJ, Poulter M, Owen F, Terwilliger JD, Westaway D, Ott J, Prusiner SB (1989a) Linkage of a prion protein missense variant to Gerstmann-Sträussler syndrome. Nature 338: 342–345

Hsiao KK, Doh-ura K, Kitamoto T, Tateishi J, Prusiner SB (1989b) A prion protein amino acid substitution in ataxic Gerstmann-Sträussler syndrome. Ann Neurol 26: 137

Hsiao KK, Scott M, Foster D, Groth DF, DeArmond SJ, Prusiner SB (1990) Spontaneous neurodegeneration in transgenic mice with mutant prion protein. Science 250: 1587–1590

Hsiao K, Meiner Z, Kahana E, Cass C, Kahana I, Avrahami D, Scarlato G, Abramsky O, Prusiner SB, Gabizon R (1991) Mutation of the prion protein in Libyan Jews with Creutzfeldt-Jakob disease. N Engl J Med 324: 1091–1097

Hsiao K, Dlouhy S, Farlow MR, Cass C, Da Costa M, Conneally M, Hodes ME, Ghetti B, Prusiner SB (1992) Mutant prion proteins in Gerstmann-Sträussler-Scheinker disease with neurofibrillary tangles. Nat Genet 1: 68–71

Hsiao KK, Groth D, Scott M, Yang S-L, Serban H, Rapp D, Foster D, Torchia M, DeArmond SJ, Prusiner SB (1994) Serial transmission in rodents of neurodegeneration from transgenic mice expressing mutant prion protein. Proc Natl Sci USA 91: 9126–9130

Huang Z, Gabriel J-M, Baldwin MA, Fletterick RJ, Prusiner SB, Cohen FE (1994) Proposed three-dimensional structure for the cellular prion protein. Proc Natl Acad Sci USA 91: 7139–7143

Jendroska K, Heinzel FP, Torchia M, Stowring L, Kretzschmar HA, Kon A, Stern A, Prusiner SB, DeArmond SJ (1991) Proteinase-resistant prion protein accumulation in Syrian hamster brain correlates with regional pathology and scrapie infectivity. Neurology 41: 1482–1490

Kitamoto T, Iizuka R, Tateishi J (1993a) An amber mutation of prion protein in Gerstmann-Sträussler syndrome with mutant PrP plaques. Biochem Biophys Res Commun 192: 525–531

Kitamoto T, Ohta M, Doh-ura K, Hitoshi S, Terao Y, Tateishi J (1993b) Novel missense variants of prion protein in Creutzfeldt-Jakob disease or Gerstmann-Sträussler syndrome. Biochem Biophys Res Commun 191: 709–714

Kretzschmar HA, Prusiner SB, Stowring LE, DeArmond SJ (1986a) Scrapie prion proteins are synthesized in neurons. Am J Pathol 122: 1–5

Kretzschmar HA, Stowring LE, Westaway D, Stubblebine WH, Prusiner SB, DeArmond SJ (1986b) Molecular cloning of a human prion protein cDNA. DNA 5: 315–324

Laplanche J-L, Chatelain J, Launay J-M, Gazengel C, Vidaud M (1990) Deletion in prion protein gene in a Moroccan family. Nucleic Acids Res 18: 6745

McKinley MP, Taraboulos A, Kenaga L, Serban D, DeArmond SJ, Stieber A, Prusiner SB (1990) Ultrastructural localization of scrapie prion proteins in secondary lysosomes of infected cultured cells. J Cell Biol 111: 316a

Medori R, Montagna P, Tritschler HJ, LeBlanc A, Cortelli P, Tinuper P, Lugaresi E, Gambetti P (1992a) Fatal familial insomnia: a second kindred with mutation of prion protein gene at codon 178. Neurology 42: 669–670

Medori R, Tritschler H-J, LeBlanc A, Villare F, Manetto V, Chen HY, Xue R, Leal S, Montagna P, Cortelli P, Tinuper P, Avoni P, Mochi M, Baruzzi A, Hauw JJ, Ott J, Lugaresi E, Autilio-Gambetti L, Gambetti P (1992b) Fatal familial insomnia, a prion disease with a mutation at codon 178 of the prion protein gene. N Engl J Med 326: 444–449

Miyazono M, Kitamoto T, Doh-ura K, Iwaki T, Tateishi J (1992) Creutzfeldt-Jakob disease with codon 129 polymorphism (Valine): a comparative study of patients with codon 102 point mutation or without mutations. Acta Neuropathol (Berl) 84: 349–354

Muramoto T, Kitamoto T, Tateishi J, Goto I (1992) The sequential development of abnormal prion protein accumulation in mice with Creutzfeldt-Jakob disease. Am J Pathol 140: 1411–1420

Nochlin D, Sumi SM, Bird TD, Snow AD, Leventhal CM, Beyreuther K, Masters CL (1989) Familial dementia with PrP-positive amyloid plaques: a variant of Gerstmann-Sträussler syndrome. Neurology 39: 910–918

Owen F, Poulter M, Lofthouse R, Collinge J, Crow TJ, Risby D, Baker HF, Ridley RM, Hsiao K, Prusiner SB (1989) Insertion in prion protein gene in familial Creutzfeldt-Jakob disease. Lancet 1: 51–52

Owen F, Poulter M, Shah T, Collinge J, Lofthouse R, Baker H, Ridley R, McVey J, Crow T (1990) An in-frame insertion in the prion protein gene in familial Creutzfeldt-Jakob disease. Mol Brain Res 7: 273–276

Palmer MS, Mahal SP, Campbell TA, Hill AF, Sidle KCL, Laplanche J-L, Collinge J (1993) Deletions in the prion protein gene are not associated with CJD. Hum Mol Genet 2: 541–544

Pan K-M, Baldwin M, Nguyen J, Gasset M, Serban A, Groth D, Mehlhorn I, Huang Z, Fletterick RJ, Cohen FE, Prusiner SB (1993) Conversion of α-helices into β-sheets features in the formation of the scrapie prion proteins. Proc Natl Acad Sci USA 90: 10962–10966

Petersen RB, Tabaton M, Berg L, Schrank B, Torack RM, Leal S, Julien J, Vital C, Deleplanque B, Pendlebury WW, Drachman D, Smith TW, Martin JJ, Oda M, Montagna P, Ott J, Autilio-Gambetti L, Lugaresi E, Gambetti P (1992) Analysis of the prion protein gene in thalamic dementia. Neurology 42: 1859–1863

Poulter M, Baker HF, Frith CD, Leach M, Lofthouse R, Ridley RM, Sah T, Owen F, Collinge J, Brown G, Hardy J, Mullan MJ, Harding AE, Bennett C, Doshi R, Crow TJ (1992) Inherited prion disease with 144 base pair gene insertion. I. Genealogical and molecular studies. Brain 115: 675–685

Prusiner SB, Bolton DC, Groth DF, Bowman KA, Cochran SP, McKinley MP (1982) Further purification and characterization of scrapie prions. Biochemistry 21: 6942–6950

Prusiner SB, McKinley MP, Bowman KA, Bolton DC, Bendheim PE, Groth DF, Glenner GG (1983) Scrapie prions aggregate to form amyloid-like birefringent rods. Cell 35: 349–358

Prusiner SB, Scott M, Foster D, Pan K-M, Groth D, Mirenda C, Torchia M, Yang S-L, Serban D, Carlson GA, Hoppe PC, Westaway D, DeArmond SJ (1990) Transgenetic studies implicate interactions between homologous PrP isoforms in scrapie prion replication. Cell 63: 673–686

Prusiner SB, Groth D, Serban A, Koehler R, Foster D, Torchia M, Burton D, Yang S-L, DeArmond SJ (1993) Ablation of the prion protein (PrP) gene in mice prevents scrapie and facilitates production of anti-PrP antibodies. Proc Natl Acad Sci USA 90: 10608–10612

Prusiner SB, Baldwin M, Collinge J, DeArmond SJ, Marsh R, Tateishi J, Weissmann C (to be published) Classification and nomenclature of viruses: prions. Arch Virol

Puckett C, Concannon P, Casey C, Hood L (1991) Genomic structure of the human prion protein gene. Am J Hum Genet 49: 320–329

Ripoll L, Laplanche J-L, Salzmann M, Jouvet A, Planques B, Dussaucy M, Chatelain J, Beaudry P, Launay J-M (1993) A new point mutation in the prion protein gene at codon 210 in Creutzfeldt-Jakob disease. Neurology 43: 1934–1938

Scott M, Foster D, Mirenda C, Serban D, Coufal F, Wälchli M, Torchia M, Groth D, Carlson G, DeArmond SJ, Westaway D, Prusiner SB (1989) Transgenic mice expressing hamster prion protein produce species-specific scrapie infectivity and amyloid plaques. Cell 59: 847–857

Scott M, Groth D, Foster D, Torchia M, Yang S-L, DeArmond SJ, Prusiner SB (1993) Propagation of prions with artificial properties in transgenic mice expressing chimeric PrP genes. Cell 73: 979–988

Shyng S-L, Heuser JE, Harris DA (1994) A glycolipid-anchored prion protein is endocytosed via clathrin-coated pits. J Cell Biol 125: 1239–1250

Stahl N, Baldwin MA, Teplow DB, Hood L, Gibson BW, Burlingame AL, Prusiner SB (1993) Structural analysis of the scrapie prion protein using mass spectrometry and amino acid sequencing. Biochemistry 32: 1991–2002

Taraboulos A, Jendroska K, Serban D, Yang S-L, DeArmond SJ, Prusiner SB (1992) Regional mapping of prion proteins in brains. Proc Natl Acad Sci USA 89: 7620–7624

Taraboulos A, Scott M, Semenov A, Avrahami D, Laszlo L, Prusiner SB (1995) Cholesterol depletion and modification of C-terminal targeting sequence of the prion protein inhibit formation of the scrapie isoform. J Cell Biol (in press)

Telling GC, Scott M, Hsiao KK, Foster D, Yang S-L, Torchia M, Sidle KCL, Collinge J, DeArmond SJ, Prusiner SB (1994) Transmission of Creutzfeldt-Jakob disease from humans to transgenic mice expressing chimeric human-mouse prion protein. Proc Natl Acad Sci USA 91: 9936–9940

Vnencak-Jones CL, Phillips JA (1992) Identification of heterogeneous PrP gene deletions in controls by detection of allele-specific heteroduplexes (DASH). Am J Hum Genet 50: 871–872

Westaway D, Goodman PA, Mirenda CA, McKinley MP, Carlson GA, Prusiner SB (1987) Distinct prion proteins in short and long scrapie incubation period mice. Cell 51: 651–662

Westaway D, Mirenda CA, Foster D, Zebarjadian Y, Scott M, Torchia M, Yang S-L, Serban H, DeArmond SJ, Ebeling C, Prusiner SB, Carlson GA (1991) Paradoxical shortening of scrapie incubation times by expression of prion protein transgenes derived from long incubation period mice. Neuron 7: 59–68

Westaway D, DeArmond SJ, Cayetano-Canlas J, Groth D, Foster D, Yang S-L, Torchia M, Carlson GA, Prusiner SB (1994) Degeneration of skeletal muscle, peripheral nerves, and the central nervous system in transgenic mice overexpressing wild-type prion proteins. Cell 76: 117–129

Ying Y-S, Anderson RGW, Rothbert KG (1992) Each caveola contains multiple glycosyl-phosphatidylinositol-anchored membrane proteins. Cold Spring Harb Symp Quant Biol 57: 593–604

Evidence for Two Prions in Yeast: [URE3] and [PSI]

R.B. Wickner and D.C. Masison

1 Introduction

A prion is an infectious protein, a concept that has its origins in studies of scrapie of sheep, kuru, and Creutzfeldt-Jakob disease of humans and similar spongiform encephalopathies of other mammals (Alper et al. 1967; Prusiner 1982; Prusiner and McKinley 1987). A prion is believed to be an altered form of a normal cellular protein that may have lost its normal function but has acquired the ability to convert the normal form of the protein to this altered (prion) form. If either the

Section on Genetics of Simple Eukaryotes, National Institute of Diabetes, Digestive and Kidney Diseases, Bldg. 8, Room 207, National Institutes of Health, Bethesda, MD 20892-0830, USA

absence of the normal protein function or the accumulation of the abnormal form is detrimental to the organism, a disease (or phenotype) is seen. Although still somewhat controversial, an impressive body of evidence supporting this idea has built up, including the UV resistance of the scrapie agent, the absence in purified scrapie agent preparations of an identifiable nucleic acid genome, epidemiological studies, and studies using transgenic mice that strongly point to this idea. These studies are well summarized in the other contributions to this volume and elsewhere (BROWN and GAJDUSEK 1991; PRUSINER 1994). Most recently, the in vitro conversion of the normal form of PrP to the prion form, induced by the presence of the prion form, has been demonstrated (KOCISKO et al. 1994).

Saccharomyces cerevisiae is known to carry two dsRNA viruses, two autonomous ssRNA replicons, a host of retroviruses (retrotransposons), and it can even support the replication of Brom mosaic virus, whose normal host is a plant (Table 1). All of these infectious elements are transmitted vertically (to their asexual offspring), and the dsRNA viruses and ssRNA replicons are transmitted horizontally (to other cells) via mating. Cytoplasm transferred from cell to cell in the process of mating transmits these elements from donor to recipient. Thus, for example, the killer character, determined by a satellite dsRNA, called M, dependent on one of the dsRNA viruses, called L-A, was first identified as a non-Mendelian (or cytoplasmically inherited) genetic element by BEVAN and coworkers

Table 1. Viruses of *Saccharomyces cerevisiae*

Virus	Genome size(kb)	Encoded proteins	Group	Features
dsRNA viruses				
L-A	4.6	Major coat protein (Gag) RNA polymerase (Pol)		–1 ribosomal frameshift makes Gag-Pol
M_1, M_2, M_3, M_{28},...	1.6 – 1.8	Preprotoxin		Satellites of L-A; processing by Kex1p, Kex2p
L-BC	4.6	Major coat protein (Gag) RNA polymerase (Pol)		
ssRNA replicons				
20S RNA (= W dsRNA)	2.9	RNA polymerase		Copy number induced by N-starvation, high temperature
23S RNA (= T dsRNA)	2.5	RNA polymerase		Copy number induced by N-starvation, high temperature
Brome mosaic virus segment 3		CAT or URA3		Replicates dependent on segments 1 and 2
Retrovirus				
TY1, Ty2	5.8	Gag, Gag-Pol	*copia*	+1 ribosomal frameshift makes Gag-Pol
Ty3	5.3	Gag, Gag-Pol	*gypsy*	+1 ribosomal frameshift makes Gag-Pol
Ty4	6.3	Gag, Gag-Pol	*copia*	
Ty5		Gag, Gag-Pol	*copia*	

in 1963 (MAKOWER and BEVAN 1963). The yeast mitochondrial DNA was likewise first detected by its genetic consequences: the nonchromosomal inheritance of certain respiration deficient mutants (EPHRUSSI et al. 1949).

2 Expected Properties of a Yeast Prion

Since a prion is defined (PRUSINER 1982) as an infectious protein, a yeast prion must be a non-Mendelian genetic element, like other infectious entities of yeast. The genetic properties of a prion should clearly distinguish it from those of a non-Mendelian nucleic acid replicon, such as the mitochondrial DNA genome, the dsRNA viruses (encoding the killer trait), or a DNA plasmid such as 2μ DNA (Table 2). The expected properties of a yeast prion are:

- Infectious = non-Mendelian genetic element, passed by cytoplasmic transfer.
- If it can be cured, it can arise again spontaneously (at some low frequency).
- Dominant phenotype of prion form = recessive phenotype of chromosomal mutant in the gene encoding the normal protein, because both have no active normal protein.
- The chromosomal gene encoding the normal form is required to propagate the prion form.
- Overproducing the normal form should increase the spontaneous occurrence of the prion form.

The mitochondrial genome can be cured by growing cells in the presence of ethidium bromide. Cured cells become respiration-deficient, and respiration-competent revertants of these cured strains cannot be isolated without reintroduction of the mitochondrial genome from another cell. Similarly, the L-A dsRNA virus and its killer toxin-encoding satellite dsRNA M can be cured by growth of cells at 39°C. Once the virus is lost, it does not return without being introduced from outside the cells. In contrast, if a prion can be cured by growth under some

Table 2. Comparison of nucleic acid replicons and putative prions: relationship with chromosomal maintenance genes

Non-Mendelian element	Chromosomal maintenance gene	Phenotype affected	Phenotype of presence of the non-Mendelian element	Phenotype of mutation in chromosomal maintenance gene
Mitochondrial DNA	*PET18,MIP1*	Respiratory competence	Competent	Deficient
M dsRNA	*MAK*	Killer toxin secretion	Killer	Nonkiller
[URE3]	*URE2*	N catabolism repression	Derepressed	Derepressed
[PSI]	*SUP35*	Translational read-through	Increased read-through	Increased read-through

condition which either prevents the prion from converting the normal form of the protein to the prion form, or which actually restores the prion form of the protein to the normal form, the cured strain can again (at some low frequency) give rise to cells carrying the prion since there is nothing to prevent the prion from arising again in that strain once the curing conditions are removed. We refer to this as reversible curability.

Chromosomal genes are generally required for the propagation of any non-Mendelian element. For example, the mitochondrial DNA genome requires the mitochondrial DNA polymerase, encoded by the chromosomal *MIP1* gene for its replication (FOURY 1989). The *mip1* mutant loses the mitochondrial genome and thus is respiration-deficient, just as any other strain that lacks mitochondrial DNA. Similarly, the chromosomal *MAK3* gene encodes an *N*-acetyltransferase whose acetylation of the L-A and M virus major coat protein is necessary for viral assembly (TERCERO and WICKNER 1992). Thus, *mak3* mutants lack the L-A and M dsRNAs and are nonkillers, just as any strain lacking these viral replicons. In contrast, a prion has a special relationship with a particular chromosomal gene on which it depends for its propagation. Since the prion propagates itself by converting the normal form of the protein into its own abnormal form, the prion cannot be propagated in a cell that does not produce the normal protein. This makes the chromosomal gene encoding the prion protein (and the normal form) a gene necessary for the propagation of the prion. However, loss of function mutations of the chromosomal gene for the normal form of the protein will produce the same phenotype as the *presence* of the prion (not the same as its absence as in the mitochondrial DNA or killer examples above; see Table 2).

Since the prion form arises from the normal form by a spontaneous change, cells in which the normal form is overproduced should have an increased incidence of the prion change simply because the target for such a change is larger.

3 [URE3] is a Non-Mendelian Genetic Element Producing Derepression of Nitrogen Catabolic Enzymes

In 1971 LACROUTE isolated mutants of yeast which could take up ureidosuccinate (ureidosuccinic acid, USA) from the medium in the presence of a good nitrogen source, such as ammonia. In normal yeast strains ureidosuccinate uptake is repressed by the presence of ammonia (DRILLIEN and LACROUTE 1972). These USA$^+$ mutants included recessive mutants in the chromosomal *URE2* gene, and the dominant non-Mendelian "mutants" which were dubbed [URE3]. The phenotypes of the chromosomal *ure2* mutants and the non-Mendelian [URE3] strains were indistinguishable, and not restricted to the ability to take up ureidosuccinate (DRILLIEN et al. 1973). Both strains were, for example, derepressed for the catabolic glutamate dehydrogenase.

Michel Aigle and Lacroute then showed that the [URE3] element could not be propagated in a *ure2-1* mutant host (AIGLE and LACROUTE 1975). This was the key result that made us think that [URE3] might be a prion.

4 [URE3] has the Properties Expected of a Prion

4.1 Reversible Curing

[URE3] is curable by growing cells on rich medium containing 5 m*M* guanidinium (WICKNER 1994). However, from a cured, purified clone one can again isolate [URE3] derivatives (WICKNER 1994). This shows that [URE3] and the wild-type state (denoted [ure3]) are alternate states of some entity, not the presence and absence of a replicating nucleic acid (Fig. 1).

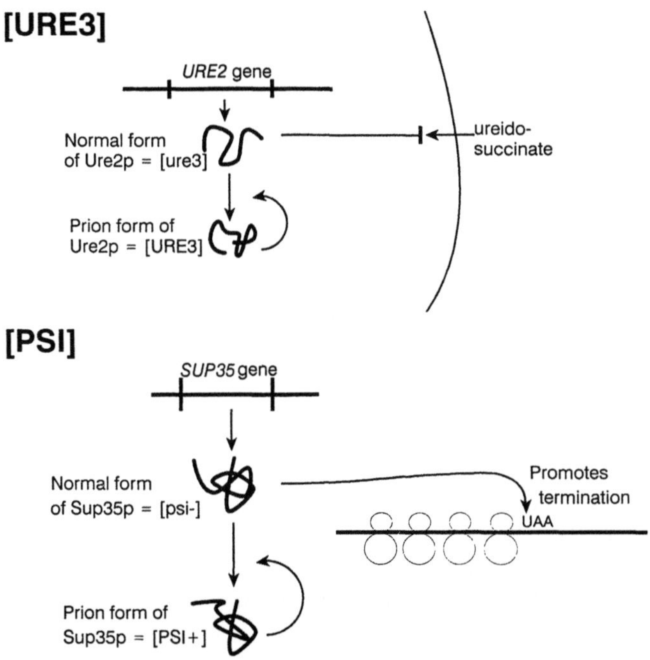

Fig. 1. Models of [URE3] and [PSI] as prion forms of Ure2p and Sup35p, respectively. Ure2p normally represses a protein needed for uptake of ureidosuccinate, as well other genes whose products are involved in nitrogen catabolism. The altered (prion) Ure2p (present in [URE3] strains) can no longer carry out the functions of the normal protein, but can now convert the normal form to the prion form. Similarly, the normal Sup35p promotes translational termination, but the prion form, found in [PSI+] strains, cannot function normally, but has acquired the ability to convert the normal form into this altered form

4.2 Dependence of [URE3] on *URE2*

Propagation of [URE3] requires the product of the chromosomal *URE2* gene, a result first reported by AIGLE and LACROUTE (1975) and confirmed by other methods by us (WICKNER 1994). What is remarkable about this result is that a gene involved in nitrogen catabolite repression (see below) is also necessary for maintaining a non-Mendelian genetic element, *and* that the phenotype of the presence of the element ([URE3]) is the same as the phenotype of the absence of this chromosomal gene (*ure2* mutation). This cannot be a coincidence.

One possible explanation of this result is that Ure2p normally acts by supporting the propagation of a nucleic acid ([ure3]) whose product is necessary for nitrogen catabolite repression. In this model [URE3] is a defective interfering (and thus dominant) mutant of [ure3] that, because it eliminates [ure3] from the cell, makes the cell defective in nitrogen catabolite repression. This model predicts that replacing the *URE2* gene would not make cells able to repress (just as replacing the *MIP1* gene in a *mip1* mutant does not bring back respiration because the mitochondrial genome is still missing). However, in fact, introduction of the *URE2* gene into a *ure2* mutant does correct the mutant defect in nitrogen catabolite repression; thus this model is wrong.

A better explanation for the relation between [URE3] and *URE2* is that [URE3] is the prion form of Ure2p, so that both [URE3] cells and *ure2* mutants lack an active Ure2p. [URE3] requires the *URE2* gene for its propagation because if no normal Ure2p is made, the prion form of Ure2p is diluted out as the cells grow and it is lost.

4.3 Overproduction of Ure2p Increases the Frequency of [URE3] Arising

Increasing the amount of the Ure2p in the cell results in a 100-fold increase in the frequency with which strains become [URE3]. This is the expected result since the development of a prion is believed to be a rare spontaneous change from the normal form. Once one molecule has changed, it presumeably catalyzes the conversion of all the normal molecules to its own form. Overproduction of Ure2p results in there being more molecules capable of the initial spontaneous change, and thus a greater likelihood of this change occurring.

While these three lines of evidence all point strongly to the prion explanation of [URE3], they do not constitute a proof of this conclusion.

5 The Role of *URE2* in Regulation of Nitrogen Metabolism

The *URE2* gene has recently been extensively characterized by Magasanik's group, and shown to negatively regulate the transcription of many genes involved in utilizing suboptimal nitrogen sources when a good nitrogen source is available. In most cases Ure2p acts with Gln3p to repress transcription (COURCHESNE and MAGASANIK 1988). Gln3p has a classic zinc finger, binds specifically to the nitrogen regulatory site upstream of at least one of the genes regulated, and appears to be a typical positive transcription regulator. The production of neither Ure2p nor Gln3 is regulated by nitrogen source, rather it is believed that an intracellular small molecule, possibly glutamine, signals the presence of an abundant nitrogen source to Ure2p which then acts on Gln3p to prevent its activation of transcription of the target genes.

N source $\rightarrow \uparrow$ glutamine $\rightarrow$ activation of Ure2p $\rightarrow$ inactivation of Gln3p $\rightarrow$ $\downarrow$transcription of target genes

Ure2p has homology with glutathione *S*-transferases, but this enzymatic activity of Ure2p, has not yet been demonstrated (COSCHIGANO and MAGASANIK 1991). It is conceivable that Ure2p acts on Gln3p by glutathionating the transcription regulator.

6 [PSI] is a Non-Mendelian Genetic Element Producing Increased Read-Through of Translation Termination Codons

[PSI] was discovered by COX in 1965 as a non-Mendelian element which elevated the efficiency of ochre suppression by an otherwise weak suppressor called SUQ5 (reviewed in COX et al. 1988; COX 1993). SUQ5 encodes a classical suppressor tRNA (LEIBMAN et al. 1975; COX 1993), and [PSI] was found to increase the efficiency of suppression of all opal and ochre suppressor tRNA mutants (UGA-and UAA-suppressing) as well as the "natural suppressors," tRNAs that can, with some low efficiency, pair with a termination codon (reviewed by COX et al. 1988). [PSI] can be lethal to a strain that carries a moderately strong suppressor as would be expected for the formation of many abnormal readthrough proteins (COX 1971).

[PSI] can be cured by exposure of cells to high osmotic strength medium (SINGH et al. 1979), or by low concentrations (1–5 m*M*) of guanidinium ion (TUITE et al. 1981). SINGH et al. (1979) showed that the curing by high osmotic strength is not due to selection for growth of cells that lacked [PSI], but that [PSI+] cells are converted to [psi–] even without substantial growth. Tuite et al. showed that this curing of [PSI] by high osmotic strength is reversible. That is, strains cured of [PSI] can again become [PSI+] without its introduction from outside of the cells. It was also reported that while curing of cells with 1 m*M* guanidine is reversible, curing with 5 m*M* is not. As is discussed below, it would be of interest to reexamine the apparently irreversibility of curing of [PSI] by 5 m*M* guanidine.

Careful studies were carried out to examine the relation of [PSI] to several known nonchromosomal replicons of *S. cerevisiae*, such as 2μ DNA, mitochondrial DNA, and the dsRNA replicons known at that time (see Table 1) and to other known non-Mendelian genetic elements (Young and Cox 1972; Tuite et al. 1982). Convincing evidence was presented that [PSI] is distinct from each of these entities. The sensitivity of [PSI] to conventional mutagens prompted a direct search for the [PSI] genome by transformation of a [psi–] strain with DNA from a [PSI+] strain. It was reported that 3μ DNA, the circular rDNA plasmid found in most yeast strains, can transmit [PSI] to a new strain (Dai et al. 1986). However further efforts to confirm this have not yet been successful (reviewed by Cox 1993).

7 [PSI] has the Expected Properties of a Prion

There is a striking parallel between the properties of [PSI] and those described above for [URE3], which led us to propose that [PSI], like [URE3], is a yeast prion (Wickner 1994; Fig. 1). As with [URE3], [PSI] can be cured by growth of cells in high osmotic strength media (Singh et al. 1979) or in the presence of low concentrations of guanidine HCl (Tuite et al. 1981). From the cured strains one can again isolate colonies carrying [PSI] (Tuite et al. 1981). The chromosomal *SUP35* gene is necessary for the propagation of [PSI] (Young and Cox 1971; Doel et al. 1994; TerAvanesyan et al. 1994) and recessive *sup35* mutants have phenotypes similar if not identical to those described for the dominant [PSI] element, presumably because both lack a normal Sup35p. Finally, overproduction of Sup35p results in a 100-fold increase in the frequency with which [psi–] strains become [PSI+] (Chernoff et al. 1993).

8 [PSI] is Dominant In Vivo and Recessive In Vitro: Further Evidence for its Prion Nature

Tuite, Cox and McLaughlin (1987) used the yeast in vitro translation system to study the effects of [PSI] on readthrough of normal translation termination codons in the presence of known suppressor tRNAs. They found that extracts of [PSI] strains catalyze efficient readthrough of amber, ochre, or opal termination codons, but that extracts of isogenic [psi–] strains do not. This was the expected result, based on the in vivo data. What was surprising were the results of the mixing experiment. When the [PSI+] and [psi–] extracts were mixed, instead of the expected dominance of the [PSI+] extract, the [psi–] extract prevailed. Addition of as little as 20% of the [psi–] extract to the [PSI+] extract completely eliminated the translational read-through by the in vitro system. This would require a very complicated explanation if [PSI] were a plasmid. One might say that [PSI] was a repressor of the synthesis of the translational read-through factor. A simpler explanation is that [PSI] is the prion form of the read-through promoting factor. Then, the results observed are the expected results. In the in vitro system conversion of the normal form (present in the [psi–] extract) to the prion form is too slow to prevent the former acting to prevent read-through. In vivo the prion form (present in the [PSI+] cell) converts the normal from to the prion (inactive) form before the assay for suppression is carried out (see Fig. 2); thus the *inactive* prion form shows dominance in vivo, but the active normal form shows dominance in vitro.

The authors showed that the factor in the [psi–] extract that promotes translation termination (inhibited read-through) is ribosome-associated and can be removed from ribosomes by a high salt wash. Sup35p shows the same type of ribosome association, further confirming that [PSI] is the prion form of Sup35p.

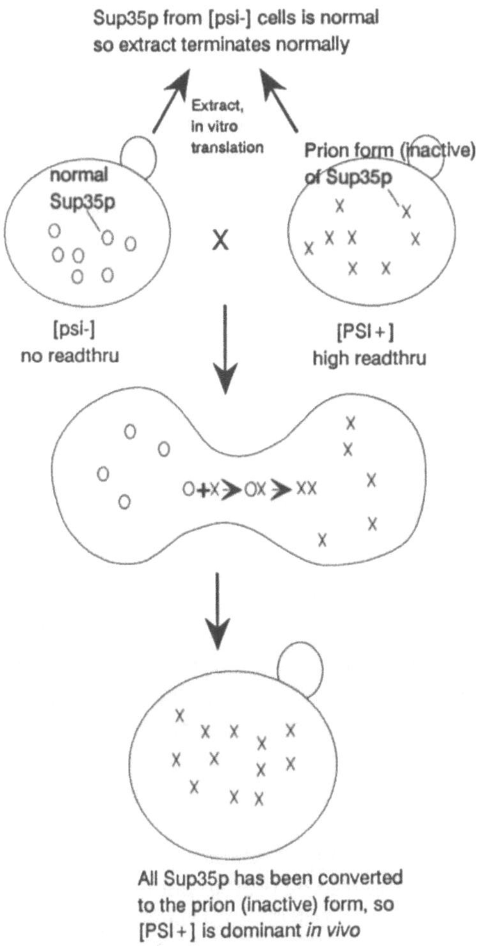

Fig. 2. [PSI+] is dominant in vivo, but recessive in vitro. We suggest that the dominance of [PSI+] in vivo results from the prion form of Sup35p in the [PSI+] parent converting the normal form in the [psi–] parent into the prion form, so that no normal form remains in the deploid cells by the time the assay is done. The in vitro conditions used by TUITE et al. (1987) for assay of translational read-through presumeably did not allow rapid conversion of normal to prion form , so the active Sup35p in the [psi-] cells showed dominance

9 Is [ETA] a Variant of [PSI] or is it a Prion Form of Another Protein?

LEIBMAN and ALL-ROBYN (1984) have described a [PSI]-like non-Mendelian genetic element, called [ETA]. [ETA] differs from [PSI] in that while [PSI] increases the efficiency of ochre suppressors, [ETA] does not. [ETA] does, however, increase the efficiency of a number of omnipotent suppressors. In addition, unlike [PSI], [ETA] is not inherited by all meiotic segregants. In this regard it resembles [URE3] (LACROUTE 1971). Chromosomal genes necessary for propagation of [ETA] have not yet been defined; it is therefore possible that [ETA] is another form of [PSI], as are the various "strains" of the scrapie prion. Alternatively, [ETA] may be a prion of another protein or not a prion at all.

10 Why so Many Prions in Yeast?

Yeast and other fungi mate very frequently in nature. For this reason yeast viruses that are passed only by cytoplasmic mixing and cannot infect by an extracellular route are nonetheless found in nearly all strains examined. This makes these organisms particularly susceptible to prions as a sexually transmitted disease. Mating in bacteria generally involves the transmission of only a strand of DNA; therefore a prion would have little chance to propagate. We expect that other prions will be found among yeasts and other organisms that exchange cytoplasm frequently.

11 What Would be Proof of the Prion Nature of [PSI] and [URE3]?

If Sup35p or Ure2p were made in *Escherichia coli* and shown to transmit the [PSI] or [URE3] traits, respectively, it could be argued that this is proof of their prion nature. However, this type of experiment would be difficult because one would not know ahead of time that these proteins were in their native state, in the prion state, or in some other conformation. Moreover, since it has already been shown that overproduction of Sup35p and Ure2p induce the conversion of wild-type strains to [PSI] and [URE3], respectively, it could be argued that one was only reproducing this result.

Of course, demonstration of biochemical changes (conformational or otherwise) of these proteins, and transmission of the trait by protein purified from the [PSI] or [URE3] strain to a [psi–] or [ure3] recipient, will be important steps.

12 Comparison of Yeast and Mammalian Prion Diseases

Alteration and accumulation of the mammalian PrP protein (Table 3) results in spongiform encephalopathy (reviewed by PRUSINER and McKINLEY 1987), but deletion of the PrP gene does not affect mouse growth or development (BUELER et al. 1992). [URE3] strains grow slowly on ammonia-containing media, perhaps because of the burden of producing many unnecessary enzymes involved in nitrogen catabolism. [PSI] strains are sick or dead because of their high levels of inappropriate readthrough of normal tanslation termination codons (COX 1971). Thus, both [URE3] and [PSI] produce disease states, but, as discussed above, the disease is due to the malfunction (or absence of function) of Ure2p and Sup35p, respectively, rather than the accumulation of the prion protein, as in mammals. Since yeast is growing exponentially, it is, of course, not expected that substantial accumulation could occur.

While the classic studies of Gajdusek on kuru, as well as those of Prusiner with the purification and transmission of the scrapie agent, have shown dramatically the infectious nature of PrP, infection by [URE3] and [PSI] have been shown only by their natural transmission route of cytoplasmic mixing. It is very important to develop an extracellular transmission assay for these elements.

Mutation of the PrP gene results in failure to propagate the scrapie agent (BUELER et al. 1993; PRUSINER et al. 1993), in the same way that mutation of *URE2* and *SUP35* result in failure to propagate [URE3] (AIGLE and LACROUTE 1975; WICKNER 1994) and [PSI] (YOUNG and COX 1971; DOEL et al. 1994; TERAVANESYAN et al. 1994). Overproduction of Ure2p and Sup35p results in increased frequency of [URE3] (WICKNER 1994) and [PSI], respectively, and apparently similar results have been seen in scrapie (WESTAWAY et al. 1994).

Table 3. Comparison of mammalian and yeast prions

Genetic property	Mammalian	[URE3]	[PSI]
Disease	Spongiform encephalopathy	N enzymes derepressed	Elevated readthrough of termination codons
Infectious	Kuru, injection	Cytoplasmic transfer	Cytoplasmic transfer
Chromosome mutation affects prion frequency	PrP gene abnormal in CJD patients	??	??
Mutation of chromosome gene with same phenotype→loss	PrPΔ →prion not propagated	ure2Δ →[URE3] not propagated	sup35 mutation → [PSI] lost
↑Normal form→ ↑ prion frequency	↑ PrP→ neurologic disease	↑ Ure2p → ↑ [URE3]	↑ Sup35p → ↑ [PSI]
Curing is reversible	??	Guanidine	Guanidine, KCl, glycerol, methanol (high osmolarity)
Altered activity or structure of protein	α-Helix →β-sheet	??	Loss of in vitro suppression enhancement

The curing of [PSI] may be true reversal of the prion change (SINGH et al. 1979). but in most cases is probably due to halting the conversion of the normal form to the prion form followed by dilution and loss of the prion form on growth of the cells. This is not yet possible in the mammalian disease.

While much is known of the structure of the mammalian prion protein and the structural changes it undergoes, there is as yet no information of this type available for the yeast prions.

13 Conclusions

Our work suggests that the prion phenomenon is not restricted to the mammalian PrP gene that causes spongiform encephalopathy when it undergoes its prion conversion. Moreover, the properties of [URE3] and [PSI] indicate that prions can be the basis of *inheritance*, as well as the basis of infection. Maternal inheritance of mitochondrial traits and viruses is known in many organisms. It is possible that maternally inherited traits may in some cases be prions. The first clear example of nonnucleic acid inheritance was the inheritance of cortical patterns in *Paramecium* described by SONNEBORN and coworkers (BEISSON and SONNEBORN 1965). It was found that microsurgically altering the patterns of the cilia on the surface of *Paramecium* resulted in a stable change that was passed on indefinitely to the offspring of surgically altered cells. The parallel of this phenomenon and the prion phenomenon has also been noted (GRIMES and AUFDERHEIDE 1991). The ease with which the yeast system can be manipulated makes [URE3] and [PSI] promising material for detailed studies of the cellular mechanisms affecting the generation, propagation, and curing of prions.

References

Aigle M, Lacroute F (1975) Genetical aspects of [URE3], a non-Mendelian, cytoplasmically inherited mutation in yeast. Mol Gen Genet 136: 327–335

Alper T, Cramp WA, Haig DA, Clarke MC (1967) Does the agent of scrapie replicate without nucleic acid? Nature 214: 764–766

Beisson J, Sonneborn TM (1965) Cytoplasmic inheritance of the organization of the cell cortex in *Paramecium aurelia*. Proc Natl Acad Sci USA 53: 275–282

Brown P, Gajdusek DC (1991) The human spongiform encephalopathies: Kuru, Creutzfeldt-Jakob disease, and the Gerstmann-Straussler-Scheinker. In: Chesebro BW (ed) Transmissible Spongiform encephalopathies: Scrapie, BSE and related human disorders. Springer, Berlin Heidelberg New York, pp 1–20 (Current topics in microbiology and immunology, Vol 172)

Bueler H, Fischer M, Lang Y et al. (1992) Normal development and behaviour of mice lacking the neuronal cell-surface PrP protein. Nature 356: 577–582

Bueler H, Aguzzi A, Sailer A et al. (1993) Mice devoid of PrP are resistant to scrapie. Cell 73: 1339–1347

Chernoff YO, Derkach IL, Inge-Vechtomov SG (1993) Multicopy SUP35 gene induces de-novo appearance of psi-like factors in the yeast *Saccharomyces cerevisiae*. Curr Genet 24: 268–270
Coschigano, PW, Magasanik B (1991) The *URE2* gene product of *Saccharomyces cerevisiae* plays an important role in the cellular response to the nitrogen source and has homology to glutathione S-transferases. Mol Cell Biol 11: 822–832
Courchesne WE, Magasanik B (1988) Regulation of nitrogen assimilation in *Saccharomyces cerevisiae*: roles of the *URE2* and *GLN3* genes. J Bacteriol. 170: 708–713
Cox BS (1965) PSI, a cytoplasmic suppressor of super-suppressor in yeast. Heredity 20: 505–521
Cox BS (1971) A recessive lethal super-suppressor mutation in yeast and other PSI phenomena. Heredity 26: 211–232
Cox BS (1993) Psi phenomena in yeast. In: Hall MN, Linder P (eds) The early days of yeast genetics. Cold Spring Harbor Laboratory Press, Cold Spring Harbor, New York, pp 219–239
Cox BS, Tuite MF, McLaughlin CS (1988) The Psi factor of yeast: a problem in inheritance. Yeast 4: 159–179
Dai H, Tsay SH, Lund PM, Cox BS (1986) Transformation of [psi–] *Saccharomyces cerevisiae* to [PSI+] with DNA co-purified with 3 µm DNA circles. Curr Genet 11: 79–82
Doel SM, McCready SJ, Nierras CR, Cox BS (1994) The dominant *PNM2* mutation which eliminates the [PSI] factor of *Saccharomyces cerevisiae* is the result of a missense mutation in the *SUP35* gene. Genetics 137: 659–670
Drillien R, Lacroute F (1972) Ureidosuccinic acid uptake in yeast and some aspects of its regulation. J Bacteriol 109: 203–208
Drillien R, Aigle M, Lacroute F (1973) Yeast mutants pleiotropically impaired in the regulation of the two glutamate dehydrogenases. Biochem. Biophys. Res. Commun 53: 367–372
Ephrussi B, Hottinguer H, Tavlitzki J (1949) Action de l'acriflavine sur les levures. II. Etude genetique du mutant "petite colonie." Ann Inst Pasteur 76: 419–450
Foury F (1989) Cloning and sequencing of the nuclear gene *MIP1* encoding the catalytic subunit of the yeast mitochondrial DNA polymerase. J Biol Chem 264: 20552–20560
Grimes GW, Aufderheide KJ (1991) Cellular aspects of pattern formation: the problem of assembly. Karger, Basel; P91
Kocisko DA, Come JH, Priola SA et al. (1994) Cell-free formation of protease-resistant prion protein. Nature 370: 471–474
Lacroute F (1971) Non-Mendelian mutation allowing ureidosuccinic acid uptake in yeast. J Bacteriol 106: 519–522
Leibman SW, All-Robyn JA (1984) A non-Mendelian factor, [eta+], causes lethality of yeast omnipotent suppressor strains. Curr Genet 8: 567–573
Leibman SW, Stewart JW, Sherman F (1975) Serine substitutions caused by an ochre suppressor in yeast. J Mol Biol 94: 595–610
Makower M, Bevan EA (1963) The inheritance of a killer character in yeast (*Saccharomyces cerevisiae*). Proc Int Congr Genet XI(1): 202
Prusiner SB (1982) Novel proteinaceous infectious particles cause scrapie. Science 216: 136–144
Prusiner SB (1994) Biology and genetics of prion diseases. Annu Rev Microbiol 48: 655–686
Prusiner SB, McKinley MP (1987) Prions. Novel infectious pathogens causing scrapie and Creutzfeldt-Jakob disease. Academic, San Diego, P 540
Prusiner SB, Groth D, Serban A et al. (1993) Ablation of the prion protein (PrP) gene in mice prevents scrapie and facilitates production of anti-PrP antibodies. Proc Natl Acad Sci USA 90: 10608–10612
Singh AC, Helms C, Sherman F (1979) Mutation of the non-Mendelian suppressor [PSI] in yeast by hypertonic media. Proc Natl Acad Sci USA 76: 1952–1956
TerAvanesyan A, Dagkesamanskaya AR, Kushnirov VV, Smirnov VN (1994) The *SUP35* omnipotent suppressor gene is involved in the maintenance of the non-Mendelian determinant [psi+] in the yeast *Saccharomyces cerevisiae*. Genetics 137: 671–676
Tercero JC, Wickner RB (1992) MAK3 encodes an N-acetyltransferase whose modification of the L-A *gag* N-terminus is necessary for virus particle assembly. J Biol Chem 267: 20277–20281
Tuite MF, Mundy CR, Cox BS (1981) Agents that cause a high frequency of genetic change from [psi+] to [psi–] in *Saccharomyces cerevisiae*. Genetics 98: 691–711
Tuite MF, Lund PM, Futcher AB, Dobson MJ, Cox BS, McLaughlin CS (1982) Relationship of the [psi] factor with other plasmids of *Saccharomyces cerevisiae*. Plasmid 8: 103–111
Tuite MF, Cox BS, McLaughlin CS (1987) A ribosome-associated inhibitor of in vitro nonsense suppression in [psi–] strains of yeast. FEBS Lett. 225: 205–208

Westaway D, DeArmond SJ, Cayetano-Canlas J et al. (1994) Degeneration of skeletal muscle, peripheral nerves, and the central nervous system in transgenic mice overexpressing wild-type prion proteins. Cell 76: 117–129

Wickner RB (1994) Evidence for a prion analog in *S. cerevisiae*: the [URE3] non-Mendelian genetic element as an altered *URE2* protein. Science 264: 566–569

Young CSH, Cox BS (1971) Extrachromosomal elements in a super-suppression system of yeast. I. A nuclear gene controlling the inheritance of the extrachromosomal elements. Heredity 26: 413–422

Young CSH, Cox BS (1972) Extrachromosomal elements in a super-suppression system of yeast. II. Relations with other extrachromosomal elements. Heredity 28: 189–199

Subject Index

Current Topics in Microbiology and Immunology

Volumes published since 1989 (and still available)

Vol. 165: **Bothwell, Mark (Ed.):** Neuronal Growth Factors. 1991. 14 figs. IX, 173 pp. ISBN 3-540-52654-4

Vol. 166: **Potter, Michael; Melchers, Fritz (Eds.):** Mechanisms in B-Cell Neoplasia. 1990. 143 figs. XIX, 380 pp. ISBN 3-540-52886-5

Vol. 167: **Kaufmann, Stefan H. E. (Ed.):** Heat Shock Proteins and Immune Response. 1991. 18 figs. IX, 214 pp. ISBN 3-540-52857-1

Vol. 168: **Mason, William S.; Seeger, Christoph (Eds.):** Hepadnaviruses. Molecular Biology and Pathogenesis. 1991. 21 figs. X, 206 pp. ISBN 3-540-53060-6

Vol. 169: **Kolakofsky, Daniel (Ed.):** Bunyaviridae. 1991. 34 figs. X, 256 pp. ISBN 3-540-53061-4

Vol. 170: **Compans, Richard W. (Ed.):** Protein Traffic in Eukaryotic Cells. Selected Reviews. 1991. 14 figs. X, 186 pp. ISBN 3-540-53631-0

Vol. 171: **Kung, Hsing-Jien; Vogt, Peter K. (Eds.):** Retroviral Insertion and Oncogene Activation. 1991. 18 figs. X, 179 pp. ISBN 3-540-53857-7

Vol. 172: **Chesebro, Bruce W. (Ed.):** Transmissible Spongiform Encephalopathies. 1991. 48 figs. X, 288 pp. ISBN 3-540-53883-6

Vol. 173: **Pfeffer, Klaus; Heeg, Klaus; Wagner, Hermann; Riethmüller, Gert (Eds.):** Function and Specificity of γ/δ TCells. 1991. 41 figs. XII, 296 pp. ISBN 3-540-53781-3

Vol. 174: **Fleischer, Bernhard; Sjögren, Hans Olov (Eds.):** Superantigens. 1991. 13 figs. IX, 137 pp. ISBN 3-540-54205-1

Vol. 175: **Aktories, Klaus (Ed.):** ADP-Ribosylating Toxins. 1992. 23 figs. IX, 148 pp. ISBN 3-540-54598-0

Vol. 176: **Holland, John J. (Ed.):** Genetic Diversity of RNA Viruses. 1992. 34 figs. IX, 226 pp. ISBN 3-540-54652-9

Vol. 177: **Müller-Sieburg, Christa; Torok-Storb, Beverly; Visser, Jan; Storb, Rainer (Eds.):** Hematopoietic Stem Cells. 1992. 18 figs. XIII, 143 pp. ISBN 3-540-54531-X

Vol. 178: **Parker, Charles J. (Ed.):** Membrane Defenses Against Attack by Complement and Perforins. 1992. 26 figs. VIII, 188 pp. ISBN 3-540-54653-7

Vol. 179: **Rouse, Barry T. (Ed.):** Herpes Simplex Virus. 1992. 9 figs. X, 180 pp. ISBN 3-540-55066-6

Vol. 180: **Sansonetti, P. J. (Ed.):** Pathogenesis of Shigellosis. 1992. 15 figs. X, 143 pp. ISBN 3-540-55058-5

Vol. 181: **Russell, Stephen W.; Gordon, Siamon (Eds.):** Macrophage Biology and Activation. 1992. 42 figs. IX, 299 pp. ISBN 3-540-55293-6

Vol. 182: **Potter, Michael; Melchers, Fritz (Eds.):** Mechanisms in B-Cell Neoplasia. 1992. 188 figs. XX, 499 pp. ISBN 3-540-55658-3

Vol. 183: **Dimmock, Nigel J.:** Neutralization of Animal Viruses. 1993. 10 figs. VII, 149 pp. ISBN 3-540-56030-0

Vol. 184: **Dunon, Dominique; Mackay, Charles R.; Imhof, Beat A. (Eds.):** Adhesion in Leukocyte Homing and Differentiation. 1993. 37 figs. IX, 260 pp. ISBN 3-540-56756-9

Vol. 185: **Ramig, Robert F. (Ed.):** Rotaviruses. 1994. 37 figs. X, 380 pp. ISBN 3-540-56761-5

Vol. 186: **zur Hausen, Harald (Ed.):** Human Pathogenic Papillomaviruses. 1994. 37 figs. XIII, 274 pp. ISBN 3-540-57193-0

Vol. 187: **Rupprecht, Charles E.; Dietzschold, Bernhard; Koprowski, Hilary (Eds.):** Lyssaviruses. 1994. 50 figs. IX, 352 pp. ISBN 3-540-57194-9

Vol. 188: **Letvin, Norman L.; Desrosiers, Ronald C. (Eds.):** Simian Immunodeficiency Virus. 1994. 37 figs. X, 240 pp. ISBN 3-540-57274-0

Vol. 189: **Oldstone, Michael B. A. (Ed.):** Cytotoxic T-Lymphocytes in Human Viral and Malaria Infections. 1994. 37 figs. IX, 210 pp. ISBN 3-540-57259-7

Vol. 190: **Koprowski, Hilary; Lipkin, W. Ian (Eds.):** Borna Disease. 1995. 33 figs. IX, 134 pp. ISBN 3-540-57388-7

Vol. 191: **ter Meulen, Volker; Billeter, Martin A. (Eds.):** Measles Virus. 1995. 23 figs. IX, 196 pp. ISBN 3-540-57389-5

Vol. 192: **Dangl, Jeffrey L. (Ed.):** Bacterial Pathogenesis of Plants and Animals. 1994. 41 figs. IX, 343 pp. ISBN 3-540-57391-7

Vol. 193: **Chen, Irvin S. Y.; Koprowski, Hilary; Srinivasan, Alagarsamy; Vogt, Peter K. (Eds.):** Transacting Functions of Human Retroviruses. 1995. 49 figs. IX, 240 pp. ISBN 3-540-57901-X

Vol. 194: **Potter, Michael; Melchers, Fritz (Eds.):** Mechanisms in B-cell Neoplasia. 1995. 152 figs. XXV, 458 pp. ISBN 3-540-58447-1

Vol. 195: **Montecucco, Cesare (Ed.):** Clostridial Neurotoxins. 1995. 28 figs. XI., 278 pp. ISBN 3-540-58452-8

Vol. 196: **Koprowski, Hilary; Maeda, Hiroshi (Eds.):** The Role of Nitric Oxide in Physiology and Pathophysiology. 1995. 21 figs. IX, 90 pp. ISBN 3-540-58214-2

Vol. 197: **Meyer, Peter (Ed.):** Gene Silencing in Higher Plants and Related Phenomena in Other Eukaryotes. 1995. 17 figs. IX, 232 pp. ISBN 3-540-58236-3

Vol. 198: **Griffiths, Gillian M.; Tschopp, Jürg (Eds.):** Pathways for Cytolysis. 1995. 45 figs. IX, 224 pp. ISBN 3-540-58725-X

Vol. 199/I: **Doerfler, Walter; Böhm, Petra (Eds.):** The Molecular Repertoire of Adenoviruses I. 1995. 51 figs. XIII, 280 pp. ISBN 3-540-58828-0

Vol. 199/II: **Doerfler, Walter; Böhm, Petra (Eds.):** The Molecular Repertoire of Adenoviruses II. 1995. 36 figs. XIII, 278 pp. ISBN 3-540-58829-9

Vol. 199/III: **Doerfler, Walter; Böhm, Petra (Eds.):** The Molecular Repertoire of Adenoviruses III. 1995. 51 figs. XIII, 310 pp. ISBN 3-540-58987-2

Vol. 200: **Kroemer, Guido; Martinez-A., Carlos (Eds.):** Apoptosis in Immunology. 1995. 14 figs. XI, 242 pp. ISBN 3-540-58756-X

Vol. 201: **Kosco-Vilbois, Marie H. (Ed.):** An Antigen Depository of the Immune System: Follicular Dendritic Cells. 1995. 39 figs. IX, 209 pp. ISBN 3-540-59013-7

Vol. 202: **Oldstone, Michael B. A.; Vitković, Ljubiša (Eds.):** HIV and Dementia. 1995. 40 figs. XIII, 279 pp. ISBN 3-540-59117-6

Vol. 203: **Sarnow, Peter (Ed.):** Cap-Independent Translation. 1995. 31 figs. XI, 183 pp. ISBN 3-540-59121-4

Vol. 204: **Saedler, Heinz; Gierl, Alfons (Eds.):** Transposable Elements. 1995. 42 figs. IX, 234 pp. ISBN 3-540-59342-X

Vol. 205: **Littman, Dan. R. (Ed.):** The CD4 Molecule. 1995. 29 figs. XIII, 182 pp. ISBN 3-540-59344-6

Vol. 206: **Chisari, Francis V.; Oldstone, Michael B. A. (Eds.):** Transgenic Models of Human Viral and Immunological Disease. 1995. 53 figs. XII, 350 pp. ISBN 3-540-59341-1

Springer-Verlag and the Environment

We at Springer-Verlag firmly believe that an international science publisher has a special obligation to the environment, and our corporate policies consistently reflect this conviction.

We also expect our business partners – paper mills, printers, packaging manufacturers, etc. – to commit themselves to using environmentally friendly materials and production processes.

The paper in this book is made from low- or no-chlorine pulp and is acid free, in conformance with international standards for paper permanency.

MIX
Papier aus verantwortungsvollen Quellen
Paper from responsible sources
FSC® C105338

If you have any concerns about our products,
you can contact us on
ProductSafety@springernature.com

In case Publisher is established outside the EU,
the EU authorized representative is:
Springer Nature Customer Service Center GmbH
Europaplatz 3, 69115 Heidelberg, Germany

Printed by Libri Plureos GmbH
in Hamburg, Germany